BIRTH OF A NEW EARTH

NEW DIRECTIONS IN CRITICAL THEORY

NEW DIRECTIONS IN CRITICAL THEORY

AMY ALLEN, GENERAL EDITOR

New Directions in Critical Theory presents outstanding classic and contemporary texts in the tradition of critical social theory, broadly construed. The series aims to renew and advance the program of critical social theory, with a particular focus on theorizing contemporary struggles around gender, race, sexuality, class, and globalization and their complex interconnections.

For a continuation of this list, see page 303.

BIRTH OF A
NEW EARTH

THE RADICAL POLITICS
OF ENVIRONMENTALISM

ADRIAN PARR

Columbia University Press   *New York*

Columbia University Press
Publishers Since 1893
New York   Chichester, West Sussex
cup.columbia.edu

Library of Congress Cataloging-in-Publication Data

Names: Parr, Adrian, author.
Title: Birth of a new earth : the radical politics of environmentalism /
Adrian Parr.
Description: New York : Columbia University Press, 2017. | Includes
bibliographical references and index.
Identifiers: LCCN 2017008305 (print) | LCCN 2017037388 (ebook) |
ISBN 9780231542456 (electronic) | ISBN 9780231180085 (cloth) | ISBN
9780231180092 (pbk.)
Subjects: LCSH: Environmentalism—Political aspects. | Green
movement—Political aspects. | Environmental degradation—Political aspects.
Classification: LCC JA75.8 (ebook) | LCC JA75.8 .P365 2017 (print) |
DDC 320.58—dc23
LC record available at https://lccn.loc.gov/2017008305

Cover design: Julia Kushnirsky
Cover image: © Katrina Graham

CONTENTS

ACKNOWLEDGMENTS

I would like to thank the Taft Research Center for a center fellowship, which allowed me to write uninterrupted for one year. The Marble House summer fellowship provided me with much-needed peace and quiet to think, read, and write in Vermont. I treasure the intellectual friendships I formed during my time at the Marble House. Many of the ideas in this book germinated during discussions with my Marble House and Taft Center fellows. Having time to think in this day and age has become a luxury, and I am truly grateful for both these fellowships.

Several people have contributed in one way or another to this project. They include Rosi Braidotti, Brad Evans, Kenneth Surin, Gustavo Esteva, Tom Conley, Slavoj Žižek, Bill McKibben, Gareth Dale, Manu Mathai, Jose Puppim de Oliveira, Santiago Zabala, Sarah Walko, Danielle Epstein, Catherine Harris, Orit Ben-Shitrit, Michael Hardt, David Schlosberg, John Meyer, Natasha Leonard, Gregg Lambert, Hélène Frichot, Frida Beckman, Susanna Heschel, Blanca Elena Jiménez-Cisneros, and Alexander Otte.

Research for several chapters was made possible by a grant from the Arts, Humanities, and Social Sciences Office of Research, University of Cincinnati.

A special thank-you to Carolyn Christov-Bakargiev for inviting me to speak at the Istanbul Biennale, where I had the good fortune to exchange ideas on the politics of aesthetics with Fernando García-Dory and test run sections of this book in a public setting.

The ideas on sustainable development would not have matured without the input of my dear friend Kadiri Galgalo. For years he has been introducing me to his community in the slums of Nairobi. I treasure the time I have spent in conversation with him and the mothers of Dagoretti. Thanks as well to the Kounkuey Design Initiative, Human Needs Project, Kenya Slum Upgrading Project, and UN-Habitat staff for sharing their work with me and for conducting site visits throughout Kibera. Sean and Jon Hughes provided a new lens through which to contemplate human autonomy and agency in the slums.

Some chapters have been previously published in peer-reviewed journals and anthologies. Chapter 1 appears in *The Oxford Handbook of Environmental Political Theory*, chapter 2 in *Green Growth: Ideology, Political Economy, and the Alternatives*, chapter 3 in *Philosophy Today*, and chapter 5 in *Theory, Culture, and Society*. I would like to thank the editors for kindly granting me permission to republish these pieces here.

Special thanks goes to Wendy Lochner at Columbia University Press for her continuing encouragement and enthusiasm for this project.

I would like to offer an additional thank-you to the staff at both the Paris and Nairobi UNESCO offices. They have been incredibly generous with their time and assistance providing me with important opportunities to receive comments from policy makers and an international community of scholars and development agencies. Several of the chapters here would not have been possible without their support and feedback.

Last but not least it takes a village to write a book. Thank you Julie, John, Sharon, Aaron, Judy, Eli, Nancy, Deborah, Lucien, Shoshana, Yehuda, and Michael. This book is for you all.

INTRODUCTION

I tell you what freedom is to me: no fear

—Nina Simone

On February 17, 2013, I participated in the Forward on Climate rally in Washington, D.C., with my seven-year-old daughter. It was quite the memorable introduction to the White House for her. We had chatted about the protest on the nine-hour drive from Cincinnati to Washington. She understood we were sending President Obama a message: reject the Keystone XL tar sands pipeline and do something serious about climate change. More than fifty thousand people attended the rally, an impressive turnout for such a cold day in the middle of winter. We warmed our bodies drumming, singing, chanting, and dancing.

That evening, my daughter pointedly asked, "Do you think the president heard us?"

To her disappointment, and mine, I replied, "So to speak."

"So to speak"—it's a curious phrase indicating there are other ways of saying the same thing. The U.S. president heard the rallying call of climate activists outside the White House . . .

so to speak. Put this way, the idiom introduces a crack in our understanding of democratic politics and the situation of the public in political life. The "so to speak" of whether or not the president had heard the protesters' demands recognizes a nondemocratic force operating amid two primary liberal signifiers— the private and public. Basically I was expressing my distrust toward the efficacy and influence of the public on democratic politics in the age of late capitalism.

The predictions of climate scientists are grave indeed, and the terrible fact of the matter is that without widespread commitment to seismically shift political thinking and action on environmental and climate policies the earth's transient climate response to the buildup of greenhouse gases in the atmosphere is estimated to be at 3.06°F (1.7°C) with values unlikely to be below 2.34°F (1.3°C).[1] Numbers fail to capture the realities of life in a warming world. Bruised bodies and splintered bones submerged in mud after a soaked mountainside collapses from days of endless rain. The panic over the rising tide as water arbitrarily gushes across the surface of the earth leaving bodies gasping, choking, convulsing for oxygen, only to eventually turn blue and fall stiff. Heat so excessive in Australia a hundred thousand bats drop dead from the sky midflight.[2] More than seventy thousand people sweated, suffocated, melted, and boiled to death during the 2003 European heat wave.[3] And if insect disease and infestation don't kill off the world's forests, then the wildfires will make up for it, moving at breakneck speed, their blue and orange flames crackling through dried-out bush, scrub, and trees and overrunning fire crews and charring the fur and skin of wildlife.[4] Or, after years of no rain people and animals desperate to quench their thirst resort to drinking a muddy liquid extracted from below a dried-up riverbed.[5] When the world becomes thirsty, life quickly gets desperate. And desperation can be a powerful catalyst for change. The question, what kind of change?

Environmental degradation and climate change produce hardships and needs people the world over have in common. Indeed the issue traverses national borders, generational divides, and even species membership. It is this feature of a deepened sense of common adversity that offers the potential to resuscitate the meaning and function of politics in contemporary human society. Regardless of environmental harms and changes in climate impacting people differently, there remains a shared human experience of hardship that will intensify as time passes. For this reason, the environmental and climate crises contain the political potential to radically change social life so it evolves into a more equitable, inclusive, collaborative, and voluntary social system. That is, as long as people and leaders embrace the revolutionary potential of the current situation.

Slavoj Žižek bluntly points out that the notion of "capitalist prosperity" is a myth, because it "cannot be universalized."[6] The wealth amassed by the world's eighty-five richest people, around US$1.7 trillion, equals the combined wealth of approximately 3.5 billion people (half the global population). A global economic system that enables the top 1 percent to enjoy sixty-five times the wealth of the bottom half of the world, around US$110 trillion, is unethical to say the least. Add into the mix the more than 6.5 million refugees, displaced because of war and persecution.[7]

The inequitable effects of neoliberal democracy, global capitalism, and militarism are alive and well. Whether we are speaking of the thousands of species forced into extinction because of the unbridled excess of human consumption and waste; the ecosystems in collapse; the nearly thirteen million hectares of global forest lost between 2000 and 2010 because of deforestation, forests that were once the world's lungs and now the source of anywhere between 12 percent and 17 percent of global greenhouse gas emissions;[8] the arctic animals starving to death during the long summer months as the sea ice melts; the changing migration patterns of birds affecting their breeding and feeding; a heat so dense and thick it blocks the lungs and slows the breath; millions living in squatter settlements and ghettos; people displaced by war endlessly waiting for refugee status to begin life over; or the homeless and hungry left to their own devices begging on street corners—the promise of freedom and choice so entrenched in everyday political jargon translates for many into "nothing left to lose." Surviving is not living.

Since the 1970s trickle-down economic theory has trickled its way into even the bluest ranks of the left wing. In a nutshell, the principles of individualism, privatization, deregulation, consumerism, and security defining global politics, economic practices, and cultural values dominate life in the twenty-first century. Social services have become profit centers; the nonprofit sector and nongovernmental organizations now operate as businesses; social and environmental responsibility agendas frame the mission statements of the corporate sector; free market environmentalism and environmental security are all the rage. I argue in this book that according to this schema the transformative politics of environmentalism is curtailed if it turns into the ideological supplement of neoliberal capitalism and militarism.

In *The Parallax View* Žižek argues Buddhism functions as the perfect ideological supplement to capitalism. Instead of dealing with the pace of social change and technological developments encountered in the contemporary world, Buddhism presents a tempting alternative: play the capitalist game with an "inner distance."[9] Let go of your material possessions, revert to a state of inner peace and silence, and take comfort in the idea that all we are really dealing with are insubstantial semblances, "an illusory theater of shadows."[10] The point here is that we fool ourselves into believing capitalism doesn't infiltrate every corner of our lives—democratic life, environmental policy, environmental politics, war, sustainable development, and collective imagination, to name a few examples addressed in this book. In so doing we facilitate the violence of virtual capitalism—speculation, finance, data, and the virtual battlefield—by not engaging with its material effects: exploitation, oppression, and suffering.

An ideological supplement is a fetish. It exemplifies the very systems and beliefs people reject. The claim that the environmental political project is turning into a fetish ought not to be confused with the idea of living in a fantasy world. Fetishists are realists. They use the fetish as a coping mechanism for reality. Latching on to the fetish is a way of dealing with reality only after the fetishist realizes how reality works. To speak of environmentalism as the ideological supplement to capitalism and militarism is to say that environmentalism under neoliberal democracy is a lie. But it is not just any old lie; it allows us to deal with an otherwise intolerable reality—the human species is on its deathbed, all the while continuing to perpetuate the problem.

Birth of a New Earth charts the three axes of environmental politics—economic life, governmentality, and militarism— identifying the moments of their intersection. I argue that the political work of environmentalism sets out to keep their intersection in perspective. The axiomatic of environmental politics, one that also serves as the source for the birth of a *new* earth, is simple: the refusal to surrender life to the violence of global capitalism, corporate governance, and militarism.

Environmental degradation refers to changes in the climate, loss of habitat and habitat fragmentation, damage inflicted on ecosystem vitality, decrease in the quality and quantity of natural resources, pollution, toxicity, and biodiversity loss. Factors influencing the degradation of the life world include rapid urbanization, industrialized agriculture, dirty energy, war, and a growing global middle class of eager consumers with unsustainable ecological footprints. These factors have crystallized in the axiomatic of capital—the production, distribution, circulation, and accumulation of capital. I argue that environmental degradation is an externalized cost of economic growth, limp and enfeebled democracies, militarization, an inequitable global society, and an apocalyptic collective imagination frozen by the horror and pleasure the spectacle of violence produces.

The opening chapter describes the rise of the environmental movement in legislation and policy. It then goes on to describe how the environmental movement shifted from being a counterpoint to capitalism to a new market opportunity. As environmentalism became increasingly popular, the radical impetus of the movement weakened. This turned the environmental problem into one of changing individual consumption patterns, and it reflected a general acquiescence toward free market solutions and corporate governance.

I contend that conflating environmentalism with economic growth is inherently political for it sanitizes the destruction capital accumulation is inflicting on the environment. Furthermore, the conflation displaces the important political issue that environmental degradation poses for society. Namely, environmental degradation presents an incredible opportunity to strategically interrupt the march of global capitalism. Mainstream environmentalism has, unfortunately, with all good intentions, participated in obscuring the deeper structural problem capitalism poses for environmental well-being.

In chapter 2 I maintain that placing environmental responsibility in the hands of government can also be problematic, especially under the increasingly pervasive model of corporate governance. Using the principles of economic rationality to respond to environmental concerns, governments are increasingly embracing the model of inclusive green growth. Yet as this chapter shows, the model assumes economic growth is politically neutral. I argue that the tripartite framework of inclusive green growth, which reconciles the struggle between the environment, social inequity, and the economic sphere, masks the social contradictions this struggle once exposed. First, I lean upon Thomas Piketty to show that economic growth is not necessarily redistributive in the way that inclusive green growth proponents assume. Indeed, by leaving the premise and rate of economic growth unchallenged, theories and practices of inclusive green growth do not adequately confront the complex political landscape of inequity, the accumulation of private wealth and property, and the logic of competition upon which capitalist market mechanisms rest. Second, I explore how inclusive green growth reproduces neoliberal governmentality, paying close attention to the constitutive effects of neoliberalism. In this regard, the ideational problem of how social relations and environmental politics become the means with which to acquire a competitive edge in a global marketplace and to produce resilient subjects is central to the machinations of what I call green governmentality.

Despite the massive infringement upon civil liberties and human rights that states of emergency have been used to justify, environmental emergencies harbor a new mutation of the philosophical problem of order versus anarchy, one where civil disobedience enters the murky waters of corporate power and anarchical rule. The failure of liberal democracies to confront the reality of their own historical excesses in concrete terms is basically what apocalyptic images of debris-covered landscapes, drowned and charred bodies, or parched and thirsty

fields present. Chapter 3 critically evaluates the figure of sovereignty as it appears in theories of emergency politics—the state
of exception, deliberative democracy, and de-exceptionalizing
the exception—arguing for a shift in theoretical focus: move
from a sovereign figure to a sovereign force as the basis of transformative politics.

So many odd couples are forming between environmentalism and reactionary politics it can be difficult to get one's political bearings. Chapter 4 describes the phenomenon of fascistic
environmentalism, explaining how some pockets of environmentalism have historically succumbed to ideological supple-
mentation as environmentalism enters the political domain. I
argue that the emancipatory promise of environmental politics
rests upon *not* deepening social, economic, cultural, and political
divisions. Relying on the power of the people's voice, the environmental movement conjures up the core principle driving
democratic theory: rule by the people.[11] The power of the people,
however, is not inherently emancipatory.

In chapter 5 I explore how the rural-metropolitan-wilderness hybrid prevalent in shrinking cities offers an emancipatory
development path for underprivileged communities. In effect,
urban farming in shrinking cities is a process of urban commoning. Urban commoning is understood as practices that
combine different environmental, economic, and social struggles. Urban commoning subverts relations of power, forcing
collective energies into the production, realization, and circulation of surplus value.

One of the most debilitating consequences of climate change
and environmental degradation is unsafe and unreliable access
to clean drinking water and healthy, affordable food. This problem is especially pronounced in slum communities throughout
the world, with women bearing the brunt of the burden when it

comes to collecting, storing, and managing domestic food and water supplies. Chapter 6 describes how design thinking and public interest design projects provide experiments in economic democratization and the democratization of health and well-being.

What is at stake in chapter 7 is the destruction of an important human habitat—the city—and the subsequent claim that this is no different from logging other significant wildlife habitats such as the Amazon forest. I argue that the leveling of cities, such as Aleppo or Gaza, is an instance of urban clear-cutting because it uniformly removes entire human habitats. Just as clear-cutting a forest destroys the habitats of wildlife, so too does the erasure of an urban environment. In war the environment is used as a weapon of war. Reducing the built environment to a pile of debris and rubble swiftly eliminates where people live and the services they rely upon to survive. To date, dealing with the debilitating ramifications of war is largely the political terrain of peace activists, humanitarian aid workers, and human rights advocates. It is not the domain of environmental politics.

The failure to recognize the effect war has on the built environment as an environmental problem marks a serious limit in environmental politics. And the limit is not restricted to simply expanding the type of issues environmentalism brings into the fold of its list of causes. Rather, it underscores a blind spot in environmental politics and the collective imagination, one that comes at the expense of a critical politics committed to taking on the structure of violence underlying the world's environmental challenges and imagining alternative futures, futures built upon solidarities formed across ethnicity, race, gender, class, generations, and species.

Imagining is inherently political because it is constituted by and in turn constitutes change. All change is an invention of sorts. The irony is that if we continue with business as usual, people will no longer need to imagine the end of the world. The human species, and our imaginations along with it, will have joined the mounting tally of species extinction. The spectacle of natural disasters indulges the senses, immobilizes memory, and overwhelms critical thinking. Chapter 8 argues apocalyptic imagination arises when suffering collapses into ecstasy. An apocalyptic imagination occurs through instantaneous fragments of time, such that the horror before us is decontextualized and stripped of history, leaving in its wake a past drained of promises and a future deprived of hope.[12] Countering the inward-looking echo chamber of apocalyptic imagination, I maintain that emancipatory imagining is adept at aligning difference and solidarity.

Birth of a New Earth is concerned with how environmental degradation redefines political ideas of solidarity and freedom, and how environmental harms challenge the way people understand and practice freedom, equality, and dignity. While some theorists and activists champion reform and consensus, others advocate for oppositional political practices, calling upon affirmative and negative politics, respectively. That is the primary paradox this book explores. It ends with the proposition, sometimes it is both.

BIRTH OF A
NEW EARTH

1

VARYING SHADES OF GREEN

Neoliberal principles of individualism, privatization, consumption, and unconstrained choice underpinning advanced capitalism are rapidly becoming the predominant strategy used in response to widespread environmental degradation and climate change. Whether we are speaking of changing our personal choices to buy green or investing in the growing market in carbon offsets, there is very little difference between the two. Both assume we can leave pollution, species extinction, the increasing frequency and intensity of extreme weather events, diminishing water quality and quantity, food scarcity, trash, and dirty energy up to the market to solve.

By not challenging the neoliberal principles of individualism, privatization, consumption, and unconstrained choice underpinning advanced capitalism, the environmental movement leaves unquestioned the unsustainable model of growth capitalism relies upon. Conflating environmentalism with neoliberalism displaces the important political issues environmental degradation poses for society. Namely, environmental degradation presents an incredible opportunity to strategically interrupt the violent march of global capitalism and the model of endless economic growth it is premised upon.

The World Bank succinctly clarifies the situation: "World trade has exploded since the early 1960s. World exports have grown just under $1 trillion a year (in 2000 dollars) to nearly $10 trillion a year, annualized growth of some 5.5 percent per year . . . They are clearly outpacing global output, which increased at some 3.1 percent per year over the same period. Between 1970 and 2004, the share of exports relative to global output has more than doubled and is now greater than 25 percent. Throughout the early part of this period the export elasticity (the rate of growth of exports relative to output) was running at about 1.5, but around 1986 the elasticity picked up substantially, peaking at more than 2.5 a decade later."[1] Economic growth has led to rapid urbanization, industrialized agriculture, dirty energy, and a growing middle class of eager consumers with unsustainable ecological footprints.[2] The environmental implications of economic growth at this scale are massive: climate change, loss of habitat, habitat fragmentation, deteriorating ecosystems, diminishing quality and quantity of natural resources, pollution, toxicity, and biodiversity loss.

Since the beginning of industrialization the environment has borne the brunt of the economic development burden providing the raw materials necessary for production and a dumping ground for pollution. Capitalism, as John Bellamy Foster, Brett Clark, and Richard York have summed it up, is "incurring an enormous ecological debt" using up "environmental resources and the absorptive capacity of the environment while displacing the costs back on Earth itself."[3] James O'Connor has called this the "second contradiction of capital."[4] That is, capitalism fails to replenish the conditions of production and in so doing underproduction sets in as the well-being of human labor and the vitality of ecological systems degrade. All in all, the direct cost of economic growth to the environment and the often indirect costs to communities living in the vicinity of environmental hazards are not factored into the overall cost of a commodity or service. In other words, environmental degradation is a negative externality of capital.

Let's take a moment to sketch some of the defining features of capital by returning to that eminent and influential thinker Karl Marx. The linchpin of Marx's understanding of capital is that it is a process, not a thing.[5] Marx explains that the process of capital accumulation consists of production, exchange, circulation, and the generation of wealth (money, assets, rent). Commodities are produced for exchange on the market to generate surplus value. They hold both a use and exchange value. A use value is directly consumed and satisfies a human need, whereas the commodity's exchange value is actualized when the commodity is sold. Marx used the formula M-C-M to summarize the process of using money to buy a commodity that is then resold for a profit.[6] Surplus value isn't realized until the commodity is sold for money. Prior to money entering the equation, surplus value exists merely as an unrealized speculative value. In this respect money doesn't just measure value it is also a medium of circulation. It is the movement arising from exchange that facilitates the accumulation of capital, because through exchange, surplus is generated.

Price does not accurately represent the real costs of a commodity. There are a range of social and environmental damages that are part of the production, circulation, and exchange of commodities that are used for free. Neither producers, nor consumers, pay for negative environmental costs such as air and water pollution, aquifer depletion, or deforestation. Some environmental costs, such as the natural resources used in production, do enter the price structure of a commodity as raw materials, but there are many environmental costs, such as emitting carbon into the atmosphere or water contamination and depletion, that come free. Furthermore, the complex environmental costs arising from each stage of a commodity's global life cycle remain externalized.[7] For example, the cost of a car does not honestly represent the environmental damage that comes from mining the coal to produce the energy needed in the production of steel or the chemicals used in plastics or spray paint, which contaminate water supplies, nor does it reflect the social and environmental costs of oil spills from the gas used to power the trucks that transport cars to dealerships. Then there are future environmental costs such as the amount of greenhouse gases (GHG) the car will pump into the atmosphere during its lifetime and that are driving climate change or the expanding number of landfills dotted around the globe that the car will eventually be dumped at. Basically producers and consumers are free riding (depleting a public good or service while reaping the advantages without paying for it) the benefits of the environment. Indeed, a report issued by the Corporate Eco Forum in 2013 estimated that nature provides "\$72 trillion worth of free goods and services" to the global economy.[8] All in all, the direct and indirect costs of production and economic growth are not included in exchange value; as a result, it is the environment and communities that end up absorbing these costs.

Capital engenders numerous negative environmental externalities. Carbon emissions from human activities are causing global climate change. According to ice core records, the earth's average concentration of carbon dioxide preindustrialization is estimated at 280 ppm.[9] In May 2013 the National Oceanic and Atmospheric Administration's Mauna Loa Observatory reported a daily average global concentration of carbon dioxide of 400 ppm.[10] To avoid significant changes to the climate, scientists advise global carbon dioxide concentrations do not exceed 350 ppm. One of the spin-off effects of climate change is the disruption of the hydrologic cycle through drought, growing frequency and intensity of storms, rising sea levels, and glacial melt. In addition to climate change, the unsustainable extraction rate of groundwater supplies is placing global water resources under stress.[11] Then there is the contamination of water supplies and the havoc this wreaks on the ecosystems that rely on healthy water. Environmental accidents such as the BP Deepwater Horizon oil spill on April 20, 2010, one of the worst oil spills in U.S. history, spewed approximately 170 million gallons of oil into the Gulf, killing eleven workers and injuring sixteen more, with severe effects for flora and fauna and injuring thousands of birds, marine life, and sea turtles.[12]

The combination of pollution, climate change, and the destruction of natural habitats has caused the planet to experience its sixth mass extinction of plants and animals, with the rate of species extinction estimated as falling between one thousand and ten thousand the natural rate by the beginning of the twenty-first century.[13] The extinction figure is an alarming indication of significant biodiversity loss, which threatens to irreparably weaken genetic diversity and in turn trigger extensive ecosystem distress and collapse.

One way to curb negative environmental externalities is through governmental regulation. The environmental movement of the 1960s and 1970s was an important ingredient in achieving this. It raised the public's environmental consciousness and placed pressure on governments to introduce environmental legislation and legal frameworks to hold polluters accountable. The environmental movement of those decades harnessed the social energies invested in other political movements—antiwar, antinuclear, labor, civil rights, and feminist—placing them in the service of an environmental agenda. It presented a unified political front traversing the differentiated landscape of identity politics popular at the time.

Celebrated environmental achievements in the United States were the Clean Air Act of 1970 and the Clean Water Act of 1972. In the United Kingdom the Clean Air Act was passed in 1968. In Europe, Norway introduced the Nature Conservation Act (1970), the European Council declared the first five-year Environment Action Programme (1972), and the Environment Protection Act was approved in Switzerland by popular vote (it didn't come into effect until 1985). In the Southern Hemisphere New Zealand approved its Clean Air Act (1972) and a Marine Reserves Act (1971), while Australia introduced the National Parks and Wildlife Conservation Act (1975), and the Australian state of New South Wales passed the Environment and Planning Assessment Act (1979). Numerous political parties with environmental agendas popped up: the Popular Movement for the Environment in Switzerland (1972); the United Kingdom's People Party (1973), which later became the Ecology Party (1975) and then the Green Party (1985); and the German Green Party (1980). The rising popularity of environmental politics ushered in numerous nongovernmental activist organizations. A few notable examples include Greenpeace (1972), which is committed to promoting peace through environmental protection and conservation, and the fund-raising activities of the World Wildlife Fund (1961), aimed at preserving the diversity of life on earth.

Why is it that despite all the achievements of the environmental movement toward the latter part of the twentieth century environmental damage continued at a brisk pace? On environmental issues, regulation and legal frameworks work only if governments serve as arbitrators between public and private interests. With the rise of neoliberalism the public role of government was seriously compromised.

Ronald Reagan may have added thousands of acres to California's state parks while governor of California, but when his time as the fortieth president of the United States ended in 1989, the environmental legacy he left was horrendous. While in office he issued leases that opened up millions of acres of national land for oil, gas, and coal development. He rolled back many environmental safeguards and set about introducing market mechanisms into environmental regulations.[14] Rolling back on government regulations meant downsizing the public sector; in this vein he cut the budget of the Environmental Protection Agency (EPA) and slowed down its enforcement program such that by 1983 EPA officials resigned en masse. In 1987 Reagan vetoed the reauthorization of the Clean Water Act, only to have this overturned by Congress. Questioning the scientific findings linking pollution to acid rain, Reagan reneged on his promise to Canadian prime minister Pierre Trudeau to honor an agreement his predecessor (President Carter) had negotiated to enforce pollution standards in the United States, pollution that had been linked to acid rain in Canada.

Crossing the oceans to the United Kingdom, another infamous proponent of neoliberalism, Prime Minister Margaret Thatcher (1979–1990), also embarked on a series of cavalier deregulatory initiatives. Thatcher refused to commit to the 30% Club of countries dedicated to reducing pollution by 30 percent on 1980 levels by 1993. (After 1988, as the condition of U.K. forests was in steady decline and pressure mounted from her conservative constituency, which had strong conservationist interests, she changed her tune.) With the Big Bang reforms of 1986 Thatcher reorganized the City of London, deregulating financial markets, turning the city into an international financial center. She instituted a private market in land by repealing the Community Land Act, which had achieved "positive planning through public control over development."[15] And she pioneered the introduction of enterprise zones that exempted developers from paying rates in an effort to liberate "land and property markets from the detailed process of development controls."[16] Enterprise zones facilitated the private development of land. Greenbelt areas were opened up for development, greater freedoms were accorded industrial and commercial developers with more "permissive attitudes to the development of large-scale out-of-town retail stores," and local authorities were "urged to adopt liberal attitudes towards development even in locations hitherto restricted for environmental reasons."[17] Furthermore, large farming interests were given preference over conservation efforts.

Basically, environmental legislation may be one way to force companies to stop the problem of free riding and internalize the costs of negative environmental externalities in capital, but this hinges upon governments' introducing environmental legislation and regulation to stop environmental harms from occurring and by bringing the full force of the law to bear on dealing with environmental harms. Neoliberal governance exacerbates the tension between advancing the public good and a state that facilitates and promotes the free market economy alongside the interests of the private sector, thereby undermining the regulatory mechanism that forces previously externalized costs to be internalized.

On the international stage, neoliberal governance resulted in financial organizations, such as the International Monetary Fund (IMF) and World Bank, imposing structural adjustment policies on "underdeveloped" countries. Policies of the IMF and World Bank boosted the power of the corporate sector and free market. Restructuring involved fiscal austerity measures, opening up the local economy to foreign investors, privatizing state-owned industries, introducing export-driven economic strategies, and deregulating local currencies. Putting the appalling abuse of human rights to one side for a moment, the environmental consequences of these restructuring efforts were frightful.[18]

Restructuring measures resulted in environmental protections and regulations being loosened or lifted entirely. For instance, under the IMF program Ghana restructured its forestry sector, replacing the Ghana Timber Marketing Board with the Timber Export Development Board to facilitate timber exports and the Forest Products Inspection Bureau to oversee production. In addition, Ghana lifted forest conservation restrictions.[19] Natural resource extraction increased to meet the demands of the export market in raw materials. The devaluation of national currencies caused prices on wood products to fall on global markets, spurring on deforestation as demand intensified. Unsurprising, during 1983 to 1993 the volume of logs exported from Ghana jumped 806 percent, and lumber exports increased 500 percent.[20] In addition, environmental goods and services began to be privatized. For example, between 1990 and 2002 the World Bank made water reform policies a condition of approximately one-third of the loans it granted, forcing lendees to agree to the privatization of the country's water resources.[21]

On the one hand, the global economy was booming amid deregulation and the massive restructuring of economies the world over. On the other hand, the environmental situation was worsening and GHG emissions were on the rise. In 1988 the Intergovernmental Panel on Climate Change (IPCC) was formed to report on the climatic impact of GHG emissions. The first IPCC assessment report (published in 1990) confirmed that human activities were connected to GHG emissions and a higher concentration of carbon dioxide in the atmosphere, which was causing the earth's surface to warm. Admitting there were some uncertainties with their predictions, the IPCC was confident enough to publicly recommend a reduction in global GHG emissions. Climate change became a public issue, and the pressure on industry actors was mounting with grassroots organizations mobilizing their forces around divestiture, such as the Global Warming Divestiture Campaign. Industry responded by joining forces to combat the legitimacy of climate science, forming industry-sponsored think tanks and associations to lobby against environmental regulations that would inevitably hurt their bottom line. The Global Climate Coalition (GCC, 1989–2002) successfully lobbied the U.S. government to avoid mandatory emissions controls in the lead-up to the Earth Summit (Rio de Janeiro, 1992). With the financial support of the oil, coal, gas, and automobile industries the GCC campaigned hard to persuade the public that GHG emissions would not pose a threat. Indeed, the GCC went so far as to claim there would be substantial benefits to higher levels of carbon dioxide, maintaining they would improve crop production and in turn alleviate poverty. However, as the scientific evidence in support of human activities' negatively causing the climate to change mounted, the smear campaign of scientific uncertainty had become an untenable position for industry actors to hold.

On December 11, 1997, the Kyoto Protocol was adopted (to be enforced on February 16, 2005). The treaty committed countries to internationally binding emissions reductions of 5 percent below 1990 levels by the years 2008 to 2012. Industry shifted gears, pushing back against environmental regulation and threatening this would lead to job cuts and economic collapse. The GCC announced that slowing emissions by 20 percent "could reduce the US gross domestic product by 4% and cost Americans up to 1.1 million jobs annually."[22] The Competitive Enterprise Institute urged the public that the Kyoto agreement would "cripple the economy." And the Heartland Institute calculated Kyoto would "cut economic growth by 50% by the year 2050."[23] Under President Bill Clinton's administration, the United States signed but did not ratify the treaty. Later, under President George W. Bush, the United States withdrew from Kyoto entirely, announcing it would be "harming the economy and hurting American workers."[24]

The GCC vehemently opposed U.S. ratification of the Kyoto Protocol on the grounds that mandatory GHG emissions reductions would hurt individual citizens, increasing the cost of gas by fifty cents a gallon, and it would lead to massive layoffs. It spent approximately thirteen million dollars on the anti-Kyoto campaign.[25] Documents reveal that the Bush administration under secretary of state Paula Dobriansky (2001–2004) acknowledged that the United States "rejected Kyoto in part based on input from you [the Global Climate Coalition]," and she expressed deep appreciation for the active involvement of Exxon executives in formulating U.S. climate change policy.[26]

This brief trot through contemporary environmental history and policy demonstrates an absence of fair play as the private sector increasingly influenced government to water down its position on environmental protections. The corporate world reached deep into its pockets to pressure government not to introduce environmental regulations, going as far as to obstruct environmental justice. In the case of climate change and emissions regulations, the solidification of neoliberal governance beginning in the 1970s, followed by an aggressive corporate campaign to cast doubt on climate science, which morphed into warning the public of the adverse economic effects of emissions regulations, all set the environmental movement back. Developed countries such as the United States and Australia dragged their heels over ratifying the Kyoto Protocol (1997), claiming it gave developing countries an unfair economic advantage. Understandably, environmentalists responded by arguing for the importance of ratifying Kyoto to avoid catastrophic changes to the earth's climate system.

What remained unquestioned, however, was the neoliberal premise at the core of Kyoto: property rights. The policy to institute carbon emissions trading presupposes that the corporate sector owns the atmosphere: something can't be sold if it isn't already owned. Not to mention it gave wealthy countries a right to dump carbon dioxide into the earth's atmosphere by buying another's surplus if they exceeded their emissions cap. Despite all the great work of leading mainstream environmental organizations such as Greenpeace or the David Suzuki Foundation, Kyoto's underlying neoliberal logic of property ownership and the privatization of carbon emissions stemming from this remain intact.[27]

More significantly, as the mainstream environmental movement fought back against the message that GHG emissions would eventually slow economic activity, they maintained that there were many economic benefits to being at the forefront of clean energy and technological solutions to climate change. Director of Greenpeace USA Global Warming Campaign Kert Davies explained, "President Bush is wrong when he says reducing GHG emissions will hurt the US economy. Bush ignores the economic benefits of US leadership on twenty-first century energy technology."[28] This sort of argument absorbs environmental activism into a system of capital accumulation.

What we are dealing with is a change in the "dominant form of valuation," as John Bellamy Foster et al. have written. That is to say, "in our age of global ecological crisis" capitalism profits from the "destruction of the planet."[29] Whereby the "growth of natural scarcity is seen as a golden opportunity in which to further privatize the world's commons" and a new means for accumulating capital.[30] This situation has given birth to a new model of economic growth called green capitalism or green economics. Put differently, environmental degradation and climate change now offer new market opportunities, paving the way for the privatization of public assets and common-pool resources.

With the publication in 1999 of *Natural Capitalism* a fresh vision of economic growth was presented. It promised a competitive advantage and healthy profits for companies that were willing to take on the challenge of leading the way using an environmentally friendly business model of reducing, reusing, and recycling waste. The goal of natural capitalism is to place social, economic, and environmental values "on the balance sheet" so that "nothing is marginalized or externalized."[31] Proponents of natural capitalism vouch for an economic system that values all forms of capital; even ecological systems are regarded as capital.

The thesis of natural capitalism is similar to the one espoused by architect William McDonough and chemist Michael Braungart, who teamed up to write *Cradle to Cradle* (an influential text in the design world). In it they advocate that the business sector make better choices that will both improve the bottom line and be environmentally friendly. Like the natural capitalists, they assume a growing market of green consumers at the other end of the production process, consumers who demand the market provide environmentally sound choices. However, McDonough and Braungart position themselves against the natural capitalism business model of reduce, reuse, and recycle, which they claim relies on an outdated model handed down to us from the Industrial Revolution. Instead of a cradle-to-grave model, they propose business do more with less by closing the loop on waste using the formula waste = food. The idea that business is inherently bad for the environment was therefore turned on its head and replaced with a model that views business as providing nourishment for the environment.[32]

Green economics recognizes that if a value for nature's goods and services can be set, this would provide enough motivation for the private sector to shift to cleaner energy sources, pollute less, and basically start investing in green technologies and business models. In so doing, it identifies a use value for nature that is both a product of human labor and satisfies a human need, along with an exchange value that comes from selling an environmentally conscious commodity.

Indeed, the optimism surrounding the greening of the economy witnessed the emergence of an entirely new breed of capitalism: the clean green capitalism epitomized by figures such as Walmart president H. Lee Scott Jr., who led the greening of Walmart initiative, and then there is the rebranding of the multinational oil and gas company BP, previously referred to as British Petroleum, which in 2000 adopted an environmentally politically correct image of Beyond Petroleum.[33] Eventually environmental initiatives and stewardship entered the corporate responsibility programs of multinational corporations such as Coca-Cola, BP, and ExxonMobil, such that what was once a counterpoint to capitalism—the environmental movement—had become the newest and latest market opportunity.

Yet not all aspects of the green economy are involved with producing green commodities. The green economy encompasses many activities that accrue capital without necessarily producing anything new. For instance, in the carbon offset market, capital is generated by simply owning, or claiming ownership of, a forest.[34] Profits are made off having control over access to water resources. Capital is accumulated through the sale of permits to use a natural resource. These are all examples of rent-seeking activities—namely, where capital is accrued not because anything new is created but because an individual or group has claimed ownership, or monopoly control, over what is otherwise a commons (a resource openly shared by different parties—people, animals, ecosystems). Rents are basically extracted from the labor of the earth's ecosystems and the reservoir of natural resources provided by the planet. The darker underbelly of power and the undue advantage that arises from it is not sufficiently addressed by green economics or natural capitalism.

The payoff of the green economy for business was sizable. In 2005 the global market in water privatization was mushrooming, with the net income of Veolia Environment valued at US$2.58 billion, with 40 percent of its sales arising from the company's water services.[35] Yet the inequitable effects of privatization are horrendous. Commenting on water privatization in Bolivia, John Foster et al. have written, "Water prices tripled and service was eliminated to those who could not afford water."[36] Environmental harms aside, clearly the economic benefits arising from the privatization of the world's common-pool resources are not being evenly distributed.

Other examples of the growth in the green economy abound. There is the growing market in carbon trading, carbon offsets, the robust private investment sector in developing and expanding clean and green energy options through the use of solar and wind, and the rise of the green commodity like the fuel-efficient Toyota hybrid Prius. In 2008 the voluntary carbon market was worth approximately US$728.1 million.[37] In 2011 Achim Steiner, the U.N. Under-Secretary-General and UNEP executive director, reported it was "up 17% to $257 billion. A six fold increase on the 2004 figure and 93% higher than the total in 2007, the year before the world financial crisis."[38]

And still, the benefits of economic gains remain concentrated in the hands of a few wealthy individuals. Meanwhile, the rest of the world has consistently been slipping further and further into a cycle of poverty.[39] The net worth of the world's two hundred richest people was estimated at US$2.7 trillion in 2012, while the poorest 3.5 billion people in the world collectively had US$2.2 trillion.[40] Meanwhile, the income for the world's wealthiest 1 percent increased 60 percent from 1993 to 2013.[41] As Naomi Klein has bleakly pointed out, with environmental harms and changes in climate adversely affecting the poor, we face a "collective future of disaster apartheid in which survival is determined by who can afford to pay for escape . . . Perhaps part of the reason why so many of our elites, both political and corporate, are so sanguine about climate change is that they are confident they will be able to buy their way out of the worst of it."[42]

By the twenty-first century it was becoming publicly offensive for industry actors to continue looking the other way on environmentally unsound business practices, a testimony to the political density of environmental degradation. Warnings over not greening the economy laid out in the *Stern Review* are just one indication of this shift in attitude. In 2006 the *Stern Review* provided an alarming synopsis of the economic costs associated with not lowering GHG emissions: "The overall costs and risks of climate change will be equivalent to losing at least 5% of global GDP each year, now and forever. If a wider range of risks and impacts is taken into account, the estimates of damage could rise to 20% of GDP or more."[43] Big business has vigorously moved from brown to green, verifying just how far society has come in dealing with the magnitude of the environmental problem. At the same time, however, the activist corner of environmentalism has weakened. The reason why is as environmental organizations professionalized, the activist impetus of the movement was politically institutionalized and the horizontal organizational structure transformed into a hierarchical one of professional leadership and an executive board driving the decision-making processes. Membership shifted from the political activist to paid members, and the political strategizing of mass mobilization and protest activities was replaced with coalition building and lobbying elected officials.[44] The more the environmental movement popularized, the more principles of neoliberalism infiltrated it. Along with it environmental issues were gradually framed as a problem of changing individual consumption patterns, bringing along with it an unquestioned acceptance of the neoliberal logic of the invisible hand of the free market working its magic to efficiently and effectively solve environmental problems.

The cultural shift that accompanied public enthusiasm for environmentalism inaugurated the emergence of a new environmentally aware consumer, and in the process the deeper systemic issue of the unsustainable nature of endless capital accumulation faded from view.[45] To avoid this shift in focus negatively impacting the bottom line, environmental externalities were addressed using market mechanisms. This has cast green business in a win-win scenario, such that industry actors could reemerge as both rescuing the environment and bolstering economic activity. Capital has successfully overcome the limit environmental degradation has posed by appropriating it and placing it back in the service of capital accumulation.

When faced with crisis, such as the looming threat of climate change or the collapse of ecosystems, those who manage the flows of capital seize upon and transform such limits into the new frontiers of capital accumulation. As Klein has concluded, "Not so long ago, disasters were periods of social leveling, rare moments when atomized communities put divisions aside and pulled together. Increasingly, however, disasters are the opposite: they provide windows into a cruel and ruthlessly divided future in which money and race buy survival."[46] In *The Shock Doctrine* Klein chronicles how environmental disasters, like other disasters (economic and conflict), are harnessed to push through neoliberal restructuring measures. This entails displacing entire communities, further disenfranchising the already disenfranchised and seizing land for redevelopment under the guise of reconstruction. For instance, the corporate opportunism of the reconstruction process that swept through both Sri Lanka after the 2004 tsunami and New Orleans in the wake of Hurricane Katrina in 2005 resulted in "victimizing the victims, exploiting the exploited."[47] The result in Sri Lanka was that the fishing communities underwent a second "tsunami of corporate globalization," first losing their homes on the beachfront to the tsunami, and second losing out to the tourism industry. The situation was not that different in New Orleans. Privatized reconstruction efforts after the storm neither employed local labor nor invested in developing much needed public infrastructure. Rather, as government resorted to outsourcing its responsibilities, the few scraps of the public that remained in New Orleans were either gutted or privatized.

When we throw trade liberalization into the mix, the poverty-environmental nexus begins to sharpen. Trade liberalization promotes pollution-intensive industries and environmentally harmful activities to be located in the so-called developing world, where environmental standards are relatively lenient and regulations that do exist are not effectively implemented because of corruption and bribery endemic to developing countries. While Judith Dean might disagree that export-led growth in China can be connected to the escalating problem of air and water pollution in that country, others such as Christopher Weber or Boqiang Lin would beg to differ.[48] Steven Davis and his colleagues have even found that not only has the outsourcing of dirty manufacturing to China caused a growth in pollution there but also that prevailing winds are now blowing China's air pollution across the ocean to the West Coast of the United States.[49]

Environmental degradation is an effect of the unequal relations of power that drive and are driven by capitalist social relations. Put simply, some people exploit the environment and benefit from that exploitation more than others. Environmental degradation exacerbates the already skewed global landscape of wealth and poverty. A minority benefits as their wealth and power expand. Meanwhile, an exploited majority remains marginalized. Viewed this way, environmental problems are socially structured just as much as they are an expression of hierarchical capitalist social relations. A political trajectory appears as the majority collectivizes around environmental issues to articulate the trenchant systems of exploitation afflicting both people and other-than-human species. In this regard environmental degradation is both a political object and subject; it is the basis for political struggle, and at the same time unequal capitalist social relations are inherent to the problem of environmental degradation. The most important lesson to take away from this is that capital accumulation and environmental degradation are inextricably joined.

The blanket statement repeated time and again is that "human activities" are responsible for the environmental crisis. Yet it is too easy to explain environmental degradation as an effect of human activities—some are surely more culpable than others. Furthermore, we need to approach cautiously the sweeping accusation against human activities. This is a neoliberal spin on a very serious problem afflicting all life on earth, for it places the solution back in the hands of individual people to change their everyday lives, and accordingly the process of capital accumulation driving environmental degradation is obscured. Making this claim is not to naively dismiss the impact human activities are having on the environment, but, more significantly, it is to turn the spotlight onto the structure of violence mediating the squeaky-clean neoliberal image of individual freedom and boundless choice whereby the problem of environmental degradation is simply solved by placing a price on nature's resources and services or by simply making a greener lifestyle choice.

Human activities do not equally inflict damage on the environment. It is important to resist oversimplifying and thus depoliticizing what is otherwise a deeply political issue, and that means not obfuscating how the human activities in question *work*. Human relations are themselves producing and reproducing capitalist social relations, and it is important to remember that not everyone in this situation has contributed equally to environmental degradation. A quick synopsis of the unequal distribution of global GHG emissions will suffice to make the point. The twenty-five countries with the largest national GHG emissions in 2004 accounted for 83 percent of global emissions, with the United States responsible for 20.6 percent. In 2004 the United States was ranked the world's largest economy. Admittedly, by 2006 as China's economy grew, its national GHG emissions surpassed those of the United States, but its per capita emissions (5.1 metric tons of carbon dioxide) were much lower than those of the United States (19.4 metric tons), which continued to rank the highest overall in the world.[50] The reason has to do with the United States having a higher standard of living and outsourcing its dirty manufacturing to countries like China. And let us not forget that wealthy Americans and Chinese alike are equally culpable.

The environmental crisis has presented humanity with an incredible opportunity to change course, to transform how society is organized around capital accumulation and experiment with more collectivist impulses that push the boundaries of private property ownership and excessive consumption. How can a price be set on the smells, colors, and textures emanating from a field bursting into bloom as springtime comes around? What price could possibly be put on the enriching sounds of wildlife awaking in the trees and rustling through the leaves? And how might the sensations arising from sunlight and shadow playfully dancing across the forest floor ever have a monetary value? These affective dimensions of life are priceless and perhaps their affective power comes from the way in which they defy abstraction in the monetary form and the privatization of collective life.

2

GREEN GOVERNMENTALITY

Governing people is not a way to force people to do what the governor wants; it is always a versatile equilibrium, with complementarity and conflicts between techniques which assure coercion and processes through which the self is constructed or modified by himself.

—Michel Foucault, "About the Beginning of the Hermeneutics of the Self"[1]

Today, on the occasion of the 60th anniversary of the founding of the Republic of Korea, I want to put forward "Low Carbon, Green Growth" as the core of the Republic's new vision. . . . Green technology will create numerous decent jobs to tackle the problem of growth without job creation. The renewable energy industry will create several times more jobs than existing industries. In the information age, the gap between the haves and have-nots has widened. On the contrary, the gap will narrow down in the age of green growth.

—Lee Myung-bak, president of the Republic of Korea[2]

This is a time when the outlook for environmental health and well-being is troubling and the culprit for many lies squarely with the economy.[3] For many the conundrum is how to introduce an environmentalist agenda into the machinations of a global capitalist economy without slowing growth. This is also a time of deep inequities in wealth, income, education, health, and opportunity. When these two factors— social inequity and environmental harms—concomitantly become political issues, they present a seemingly untenable political reality especially for those on the Left who aspire to strengthen and increase social welfare programs. Social services, so the party line goes, depend upon solid economic growth for funding, and yet growth also causes the climate to change and environmental harms to proliferate.

The political problem becomes ostensibly more insurmountable with the realization that environmental and social injustices intersect.[4] That is, poor environmental health disproportionately affects marginalized sections of the population. The literature on environmental justice presents three political axes for analysis and critique.[5] First, representative politics has failed to fully represent the civic realm. Second, there are some serious policy omissions that need to be rectified in order to better protect the rights of all citizens and fairly distribute the costs and benefits of environmental goods and harms. Third, the law has not kept up with the times and needs to be modified and better implemented so as to prevent environmental injustices from occurring.

Organized in this way, the scaffolding of some environmental justice advocates falls into a liberal trap. (1) The institution of representative democracy needs to better integrate the interests of individual political actors. (2) Economic policy needs to use the tools of the free market and create a new market in environmental goods and services all the while incorporating environmental externalities into the price structure. (3) The legal apparatus needs to make offenders legally accountable by punishing irresponsible behavior (remembering a corporation now enjoys the legal standing of personhood). Inclusive green growth neatly ties together all three strands of these solutions.

Using the principles of economic rationality to respond to environmental concerns, inclusive green growth advocates assume economic growth is politically neutral. In this chapter I argue that the tripartite framework of inclusive green growth, which reconciles the struggle between the environment, social inequity, and the economic sphere, masks the social contradictions this struggle once exposed. First, I lean upon Thomas Piketty to show that economic growth is not necessarily redistributive in the way that inclusive green growth proponents assume.[6] Indeed, by leaving the premise and rate of economic growth unchallenged, theories and practices of inclusive green growth do not adequately confront the complex political landscape of inequity, the accumulation of private wealth and property, and the logic of competition upon which capitalist market mechanisms rest. Second, I explore how inclusive green growth reproduces neoliberal governmentality, paying close attention to the constitutive effects of neoliberalism. In this regard, the ideational problem of how social relations and environmental politics become the means with which to acquire a competitive edge in a global marketplace and to produce resilient subjects is central to the machinations of what I call green governmentality.

WHO DOES GOVERNMENT SPEAK FOR?

Inclusive green growth refers to a low carbon model of efficient economic growth that uses a combination of economic and policy instruments to lower carbon emissions, foster investment in clean energy technologies, produce environmentally friendly commodities, grow the green job market, develop and institute green infrastructure, adopt integrated natural resource management, and invest in climate change mitigation strategies, all the while aspiring to improve human well-being and social equity.[7] The problem of environmental degradation is approached as an opportunity to transform the economy and unjust social relations. The theories and practices of inclusive green growth clearly attempt to bring the principle of economic growth into better balance with social and environmental justice concerns.[8] In this respect, the model has a great deal of potential insofar as it does not artificially separate economic, social, and environmental variables; rather, it addresses all three in tandem.[9]

In 2012 the United Nations Economic and Social Commission for Asia and the Pacific, the Asian Development Bank, and the United Nations Environment Programme issued a report on *Green Growth, Resources, and Resilience.* The report advocates adopting a model of economic growth that both minimizes the negative costs to the environment and maximizes social inclusiveness by working toward a "future where all people have an opportunity for a better life." It states that the "global market for green goods and services is vast and growing fast," correlating competitive economic advantage with inclusive green growth.[10] The primary argument put forth in the report states that if a country is committed to instituting policies and strategies that can ease the environmental costs associated with meeting the demands for food, housing, transportation, water, and energy, then that country would enjoy the economic opportunities that come from transforming a dirty energy, intensive resource use economy to a low carbon efficient one. Explicitly, the authors argue this will result in more jobs, economic growth, and prosperity. More significantly, the report stresses the importance of aligning economic growth with the inclusive and accessible objectives of sustainable development.

Inclusive green growth presumes economic *growth* is an a priori social good. The assumption rests with the argument that each and every day the economy serves more and more people.[11] That is, economic growth can be neatly aligned with a rise in living standards and a fall in poverty rates. In the words of World Bank Group president Jim Yong Kim, "We live in a time of great contrasts, when fewer than 100 people control as much of the world's wealth as the poorest 3.5 billion combined. But we also live in a time when many developing countries have the strongest growth rates in the world, which each year helps millions of people lift themselves out of extreme poverty."[12] The primary policy tools used to achieve economic growth in the service of an inclusive green agenda include eco-tax reform, subsidies, green stimulus investments, investments in green infrastructure, ecotourism schemes, improved natural resource management, socially inclusive development, and research and development that addresses knowledge gaps. The suggested approach is integrative and systemic, using the principles of efficiency and maximizing market solutions. Unsurprisingly, proponents of inclusive green growth view one of the challenges to realizing significant efficiency gains as "a lack of instruments to 'monetize' the benefits of conservation and efficiency and to reward sustainable consumption."[13] The solution to this is to introduce more innovative financing tools and to substantially increase investments in economic activities that improve natural capital. Emphasis is placed on the importance of public and private sector partnerships, whereby the private sector is viewed as an "active partner of governments," in addition to the institutionalization of local knowledge and politics, which has triggered a gamut of new political actors such as nonprofit and nongovernmental organizations, communities, and religious groups entering the political frame of economic policy making.[14]

Despite ambitiously setting out to bring social, economic, and environmental concerns into balance by adopting a multiscalar approach to governance that institutionalizes local political actors, the inclusive green growth model continues to push the neoliberal idea that the role of government is to encourage economic efficiency and privatization.[15] Unsurprisingly, one of the primary strategies of green growth is the privatization of the commons. By putting a price on what are otherwise common-pool resources such as water, forests, and ecosystem services, the model puts its faith in market mechanisms to solve social and environmental issues without defusing the circuitry of structural violence driving social inequity and environmental degradation.

Environmental degradation and climate change are problems we share with people in different parts of the world, with other species, and future generations. These are therefore potentially solidarity-building issues. Although proponents of inclusive green growth acknowledge the common at the core of environmental issues, by analyzing the situation from the vantage point of economic efficiency and by using criteria such as surplus value, private property, and price, the commoning condition of the problem is appropriated and privatized. In the name of saving the commons, inclusive green growth takes the self-sufficient character of a commons and inscribes it with the law of the market, thus denying the commons its distinctly autonomous condition. Accordingly, the commons is rendered noncommon. In the process the violence capital inflicts on all life-forms, a violence environmental degradation and climate change forces out into the open, becomes invisible and inaudible once more. It is at this juncture where the political significance of prioritizing economic growth and efficiency comes into the light of day. Predictably, the assumption driving inclusive green growth is that economic growth is value neutral, that it is an unquestionable good in and of itself. However, as Piketty has demonstrated, economic growth is not a reliable predictor of justice and equality, and the reason is not just economic, it is political.

In what is nothing short of a tour de force analysis of rising global inequality, Piketty's study of global economic history reaches as far back as the late eighteenth century. The three pillars of his account of economic history are capital accumulation, population growth, and technological advances. In particular, he shows how individual capital accumulation has superseded national capital accumulation. As a few individuals enjoy the majority of wealth and income, it is fair to say that today the world is experiencing a return to the patrimonial capitalism of America's Gilded Age or Europe's Belle Époque. Europe during the late nineteenth century had a high level of private wealth, with the "total amount of private wealth hovering around six or seven years of national income."[16] After two world wars the capital-income ratio fell to two or three. However, from 1950 the figure rose such that "private fortunes in the early twenty-first century seem to be on the verge of returning to five or six years of national income in both Britain and France."[17] In the United States the "top decile claimed as much as 45–50 percent of national income in the 1910s–1920s before dropping to 30–35 percent by the end of the 1940s . . . We subsequently see a rapid rise in inequality in the 1980s, until by 2000 we have returned to a level on the order of 45–50 percent of national income."[18]

The interesting point Piketty makes is that inequality doesn't grow only because of wage disparities but, more importantly, when the rate of return on capital exceeds the rate of economic growth. What this means is that inherited wealth steadily rises more than output and income. He clarifies, "People with inherited wealth need save only a portion of their income from capital to see that capital grow more quickly than the economy as a whole. Under such conditions, it is almost inevitable that inherited wealth will dominate wealth amassed from a lifetime's labor by a wide margin, and the concentration of capital will attain extremely high levels."[19] On this issue Piketty offers an important take-home lesson: serving the interests of elites, government policies have skewed the socioeconomic landscape in a manner that favors the most financially privileged members of society.

Piketty is especially critical of the failure of economic policies to adequately redress and redistribute income. Wealth is no longer tied to the innovations of an entrepreneurial and creative class; rather, the income of the extremely rich is increasingly generated from capital more than earnings. What this means is that there is less investment back into the economy. This negatively impacts job growth. When we add the realities of a growing global population into the mix, the picture becomes bleak indeed. He describes a situation of growing economic disparity and an economy dominated by those with inherited wealth. He warns that without government's taking a proactive role in rectifying the imbalance through progressive taxation, inequality will worsen with the rate of capital return on capital being greater than the rate of economic growth, which he summarizes with the following equation: $r > g$.[20] The reason for this massive shift in economic power lies with government. He warns that in a world where economic growth "generates arbitrary and unsustainable inequalities that radically undermine the meritocratic values on which democratic societies are based," it is a tough call to assume those selfsame weakened political institutions will have enough muscle and clout to put the needs and interests of disenfranchised people ahead of the special interests that exert undue influence over governments across the world.[21]

The economic portrait Piketty paints links to David Harvey's analysis of neoliberal ideology and the process of neoliberalization, which, as Harvey insists, "has celebrated the role of the rentier, cut taxes on the rich, privileged dividends and speculative gains over wages and salaries, and unleashed untold though geographically contained financial crises, with devastating effects on employment and life chances in country after country."[22]

Generally speaking proponents of neoliberal economics aspired to disembed the market economy from the regulatory structure of the state. Notable examples include the two icons of neoliberal governance, Prime Minister Margaret Thatcher and President Ronald Reagan, whose modus operandi was to roll back on big government and restructure the economy.

Yet we would be digging our heads in the sand if we ignored the many governmental ghosts that continue to lurk in the shadows of neoliberal capitalism: governments the world over have privatized public assets, deregulated exchange rates, lowered taxes on businesses and investors, cut back on social welfare programs and turned them into services and products to be exchanged on the free market, and instituted international policies that promote global economic integration by opening borders to the movement of capital, goods, and services (concomitantly stiffening controls over human migration). When taken together, all these examples attest to the widespread "shock treatment," to borrow from Naomi Klein, governments inflict upon society, restructuring it into a launchpad for the advancement of private interests.[23] This means the problem of neoliberal governance is not simply a matter of rolling back on big government; rather, the problem is what Jamie Peck has aptly described as "market-oriented governance" and the manner in which neoliberalism has become an "adaptive form of regulatory practice," as Philip Mirowski insists.[24] It is important to ask, who does government speak for? Who benefits from governmental regulation—the powerful interest groups or the common interest?

Citing Karl Polanyi, Harvey states that the "neoliberal debasement of the concept of freedom 'into a mere advocacy of free enterprise' can only mean, as Karl Polyani points out, 'the fullness of freedom for those whose income, leisure and security need no enhancing, and a mere pittance of liberty for the people, who may in vain attempt to make use of their democratic rights to gain shelter from the power of the owners of property.'"[25] We can get a sense of how this plays out in practical terms if we follow the money trail of campaign contributions, kickbacks, lobbying, bribery, and corruption, all of which have compromised the emancipatory pulse of democratic political life in the late twentieth and early twenty-first centuries. Or, to borrow from Mirowski, a critical assessment of neoliberalism requires that analysis move into the "weeds of everyday market governance, routine regulatory failure, and unprincipled political accommodation."[26]

The lobbying power of private industries directly undermines the representative power of Western democracies. The figures compiled by the Center for Responsive Politics neatly summarizes the situation in the United States: from 1998 to 2016 the total amount spent by companies, labor unions, and other organizations on lobbying elected officials and political candidates grew from $1.45 billion to $3.12 billion.[27] In the European Union after the 2008 financial crisis, the financial industry spent more than £120 million a year lobbying against regulation of the financial industry.[28] In the United Kingdom, from 2010 to 2013 government ministers reportedly met with representatives of the big six energy giants 128 times and only 26 times with representatives of energy consumers.[29]

On the international stage of geopolitics the "sole effect" and "police powers" doctrines of international law regularly collide. For instance, when environmental concerns were raised over a Mexican waste treatment plant, the Mexican government did not renew Tecmed Corporation's operating permit for the landfill. In 2003 Tecmed responded by suing the Mexican government for US$5.5 million in compensation for damages.[30] Then there is the case of the U.S. fracking company Lone Pine Resources, which, under Article 1117 of the North American Trade Agreement (NAFTA), sued Quebec for $250 million when the government introduced a moratorium on hydraulic fracturing to study the environmental effects of fracking. Article 1117 of NAFTA states that an investor can sue on the basis of damage to a "valuable right . . . without due process, without compensation and with no cognizable public purpose."[31] The company claimed it had invested millions in acquiring the permits for natural gas exploration under the Saint Lawrence River and that those permits were abruptly revoked without consultation.

A bridging country (one between developed and developing), such as South Korea, suffers from the same political challenges posed by the financial interests of the private sector. Heavy criticism has been leveled against green growth exponent and visionary of the Global Green Growth Institute former South Korean president Lee Myung-bak for his star green growth project, the Four Rivers Project. In their detailed evaluation of the project, Bettina Bluemling and Sun-Jin Yun describe the underhanded politics of Korea's Four Rivers Project and the definition of Korea as a water-stressed country by government agencies to justify the project, despite South Korea's water resources ranking forty-third among 147 countries.[32] The project went ahead despite widespread public opposition to it, including more than ten thousand law suits against the government for violating the Rivers Act, the Cultural Heritage Protection Act, and the National Finance Act, to name a few. The Korean Board of Audit and Inspection report accused the former president of corruption during the bidding process, explaining this had inflated the cost of the project. More significantly, the project suffers from poor water quality management, with the new weirs causing algae blooms and the destruction of ecological corridors around the waterways. The problem is so acute that Park Chang-geun, professor of civil engineering at Kwandong University, recommended "dismantling the weirs and restoring the rivers to their natural state."[33] The problematic political landscape of green growth policy in South Korea that Bluemling and Yun describe comes from the assumption that the primary objective of environmental policy is economic growth at all costs, even if it comes at the expense of socially repressive outcomes and ironically despite negative ecological outcomes.

Basically, the issue at stake is the discursive operation of environmental policy geared to the realization of growth and the abuse of power, whether by governments or the private sector to facilitate this. Here it is helpful to briefly recall Piketty and his historical analysis of the data, which shows that economic growth has not produced a level playing field. The reason is we are caught in a vicious cycle of the most economically privileged members of society enjoying undue influence over governments, which in turn concentrates wealth in the hands of a few. This would lead us to the conclusion that the primary political problem inclusive green growth presents is the dominance of neoliberal political and economic practices and institutions.

Drawing this conclusion, however, misses the other half of the neoliberal story. For instance, what Piketty doesn't address as he laments the demise of a "creative class" and all the investment in job growth and a growing middle class that this supposedly brings is that the term "creative class" is not politically neutral. Mirowski elucidates how governments use "creativity" to facilitate neoliberal agendas, such as the "new urban imperative" that cities become more hip (lingo for gentrified) and attractive to function as a drawing card for the mobile and creative class so as to give cities a competitive edge alongside the global cities of the world and the dominant economic position they enjoy in a global and rapidly urbanizing economy.[34] It all comes back to the logic of competition at the core of capital accumulation.

Inclusive green growth is seen to provide a competitive advantage in what is otherwise a weak global economy. In the words of Lee Myung-bak, "If we make up our minds before others and take action we will be able to lead green growth and take the initiative in creating a new civilization" that, he envisions, will use "low-emission and climate-resilient development pathways."[35] In this scenario, good governance creates the conditions for a resilient and competitive economy. Similarly, inclusive green growth policies are being employed as a means of promoting competition between individuals and facilitating competitive social relations, which are believed to be the most efficient and effective way to motivate socioeconomic change and competitively position a state in the global economy. Indeed, the World Bank reports that "the cost of environmental degradation, estimated at 9 percent of gross domestic product (GDP), is threatening both economic competitiveness and welfare."[36]

If inclusive green growth places the subject in capitalist relations of production and the realization of surplus value (growth and competition) and neoliberal ideological relations of signification (independent, responsible, motivated economic actors), what kind of political project does inclusive green growth support? What kind of political subjectivity does inclusive green growth produce? By asking these questions, I am interested especially in examining inclusive green growth as a mode of neoliberal governance. The issue these questions share is one of political knowledge and the kind of rationality employed in governance. Moreover, I am interested in how forms of neoliberal rationality inscribe themselves in green governance practices, and what role they play in them.[37] In other words, following Foucault, inclusive green growth policy is not just another instance of the misuse of state or corporate power, it is a relational mechanism of power that is strategically exercised to inject a principle of competition into the core of social relations. This prompts me to return to the representative question concerning *who* governments speak for. The question now leaves the domain of representational politics and transforms into a machinic problem of *how* governments speak. The answer: government speaks using a neoliberal rationality whereby the logic of competition driving capital accumulation is adapted as a principle of governance. This is an ideational matter.

HOW DOES GOVERNMENT SPEAK?

Government, in the way Foucault intends, not only denotes "political structures or . . . the management of states" but also refers to how government designates—namely, "the way in which the conduct of individuals or of groups might be directed."[38] Foucault posits three types of government that align with different ways in which power has historically been exercised. First, sovereign power exercises control over land and wealth, with people obeying the law. Second, disciplinary power marks the rise of the administrative state, where individual bodies and behaviors are regulated and disciplined as they are situated in a variety of institutional arrangements (school, asylum, hospital, prison, barracks). Third, governmentality marks the advent of biopower and the management of populations through the use of statistical knowledge.

Governmentality, Foucault tells us, entails the governance of conduct, "the way in which one conducts the conduct" of people.[39] That is, government is not just the political practice of sovereign states governing a population but also operates as a system of management in other domains such as the self, family, or religious guidance. Foucault maintains the state is both configured by and in turn configures individuals. In this reflexive schema of mutual determination, Foucault breaks from the centralized understanding of power offered by liberal and Marxist theories. For the former, power is a juridico-political concept legitimating sovereign state governments, whereas Marxists argue that the bourgeoisie uses the state as a means to advance its economic interests. Foucauldian power, however, is a strategic operation involving political governance, economic management, and individual self-rule. The state is not a fixed or universal entity. It is a practice involving a variety of activities and institutional configurations that come together to produce a system of governmental rationality Foucault terms neoliberal governmentality, one of the primary effects of which is the reconfiguration of social relations around a principle of competition, a clear example of government ensuring capital functions as capital.

James Meadway has pointed out that by erasing the difference between wealth and capital, Piketty's analysis has the unfortunate result of occluding different forms of capital. If capital "is required for production over time, and if it (in practice) takes different forms—buildings, machinery, office equipment, and so on, then it cannot," he says, "be collapsed into a single measure of "wealth."[40] More worrying for Meadway is the manner in which the single measure of wealth conceals the important role competition plays in ensuring capital functions as capital. The situation of endless economic growth is the result of the owners of capital competing with one another. Like Marx, Meadway does not assume growth entails a choice. Furthermore, he perceptively draws attention to the hidden neoliberal assumption of choice that lies at the heart of market-based solutions to the environmental problem and governmental regulations aimed at curbing environmental degradation. The artificial separation of the conditions needed to create economic growth, such as the accumulation process and a falling rate of profit that leads to greater exploitation of labor and the environment, from the conditions driving growth—namely, competition—detracts from the larger issue of how best to reorganize the economy in such a way that the competitive relationships underpinning growth are removed.

Foucault observed that under liberalism market activity was premised upon equivalences established through the process of exchange between two individuals. The shift to the logic of the free market meant that, unlike in the system of exchange, the principle of exchange was replaced with that of competition. Competition establishes price, making price the organizing principle for choice. Inclusive green growth as a series of policy initiatives governments institute to promote competition and to internalize negative environmental externalities using the price mechanism reinforces Foucault's point that neoliberalism does not entail rolling back on government in the way that Friedrich Hayek viewed it.[41] Instead of pushing the public sector under the hammer of the private sector, neoliberal governmentality turns the inequities produced by the free market into an economic rather than political issue, the effect of which is to disguise the exploitation and violence produced by the slippery terrain of neoliberal institutions, practices, and reasoning that morph to fit the times and stay afloat one financial meltdown at a time. As Mirowski has astutely announced, "By its nature, as an oxymoronic form of 'market rule,' neoliberalism is contradictory and polymorphic. It will not be fixed," going on to concede, "neoliberalism, in its various guises, has always been about the capture and reuse of the state, in the interests of shaping a pro-corporate, freer trading 'market order,' even though this has never been a process of cookie cutter replication of an unproblematic strategy."[42] Inclusive green growth is just another chapter in a long but contested history of neoliberal promiscuity as it carves out new ways for governments and the corporate sector to collaborate, this time on the back of socioenvironmental politics.

Aspiring to bring the poor into the fold of the global economy, in the way that inclusive green growth advocates, detracts from the deeper issue of how in the first instance the poor are unequally incorporated into the free market economy as casualties of the principle of competition driving economic relations. Add into the mix the aims of the inclusive green growth paradigm (developing new green technologies or a free market in carbon offsets) to solve environmental harms by commercializing or putting a price on them, and we are left with a potent, albeit misplaced, gesture of inclusive governance that sets out to solve the two problems of poverty and environmental injustice in one fell swoop, whereby the source of the problem is used as the solution. Inclusive green growth extends the power and reach of the free market (competition, privatization, commodification) that establishes the unequal location of people and the exploitative situation of the environment in the global economy, thus reproducing the selfsame political rationality and system of power relations. Capital accumulation and the logic of competition driving social and environmental injustices are no longer viewed through a political lens as the cause of these injustices; rather, using an economic frame that views capital and competition as the solution to injustice simply depoliticizes the economic practice of accumulation and competition.

When the principle of competition reappears as part of another discourse around resiliency, the biopolitical character of inclusive green growth is exposed. On the one hand inclusive green growth policies and actions are endorsed as promoting more resilient economies. On the other hand, resilience "has emerged as an important reference point when discussing appropriate responses to the increasing levels of risk faced by societies and economies."[43] In this way, the model of inclusive green growth neatly aligns with the growing popularity around producing resilient subjects who are better positioned to fend off the catastrophic effects of climate change, environmental degradation, and impoverishment.[44] A resilient subject is flexible and better equipped to adapt to the variegated landscape of threats, crises, and emergencies. Brad Evans and Julian Reid have uncovered the political implications of resilience theory, skillfully describing how it places the responsibility for security squarely back in the court of the world's most vulnerable subjects, who, it should be noted, are unequally impacted by environmental harms and the cycle of economic boom and bust.[45]

Joining resilience to inclusive green growth morphs the political subjectivity of resistance into one of adaptation. Instead of a politics that demands more of our politicians and political institutions and that imagines alternatives to what currently exists, we are left with a fragmented social field of individual agents responsible for their own future in the face of a hostile reality, absorbing the multiplicity of threats that come their way. Resigned to a "catastrophic imaginary" that accepts the definition of life as one big endless disaster zone, as I discuss in chapter 8, succumbing social life to the limited political horizon that insists the only road through the rubble is to try to render subjects more secure in an increasingly precarious world only creates a deeply depoliticized subject. For it means we surrender ourselves to the prospect of living dangerously, and along with it we strip transformative politics of its affirmative and creative pulse.[46] In this way, the model of inclusive green growth constitutes a system of green governmentality.

Green governmentality, in the way I am using the term, refers to a neoliberal rationality that treats the environmental crisis as a new politics of truth, one that then legitimizes the expansion of governance into new areas of mediation, legalization, management, and supervision. One such example would be the wave of new political protagonists entering the realm of governance, such as community groups, nongovernmental organizations, nonprofits, and religious institutions. These politically informal strategies of government have gone on to reshape social relations by obscuring the difference between economic and political power. Indeed, in the World Bank's working paper on green growth, its authors clearly state, "Growth does not cause inequality—policies do."[47] This conclusion not only protects the economic sphere from criticism by clearly pointing the finger of blame in the direction of government but also rests upon a false distinction between the economic and political spheres, one that is arguably an effect of neoliberal governmentality.

Green governmentality uses a variety of policy instruments, ranging from price-based mechanisms along with "norms and regulation, public production and direct investment, information creation and dissemination, education and moral suasion."[48] The easy conflation of an economically rational view of the individual with a morally responsible one is telling indeed. The focus on priming individuals to accept green growth policies through educational outreach and "moral suasion" transfers responsibility for the risks arising from living in polluted environments and the economic hardships nonelite members of society incur as the economy slows. Although neoliberal governmentality has comfortably adapted to the once challenging demands of environmental and social justice concerns, it is nevertheless important to reflect upon how this runs along a very different current to that of the more activist strains of the environmental and social justice movements. The leitmotif of *governance* in inclusive green growth theory and practices recognizes the significant role government activities, rationalities, and practices play in reconfiguring subjects away from a collectivist political subjectivity in solidarity with others into a competitively resilient subject whose independence is achieved at the expense of another's autonomy. Green governmentality aspires to produce a competitive social configuration made up of a potpourri of resilient subjects. This simply reinstates neoliberalism as the horizon of socioenvironmental politics, rejecting alternative political agendas and experiments that form solidarities across different political lines in an effort to institute emancipatory political agendas.

For these reasons green governmentality fails to tackle the deeper political struggles ensuing between the public and private sectors. At the heart of that struggle is the problem of neoliberal governance, whereby vested interests and relations of power orient how society is organized, complemented by a logic of competition governments employ to ensure capital functions as capital. As Wendy Brown has formulated it, subjects are interpreted as economic actors. "Neoliberal reason" is the term Brown uses to capture the economization of all spheres of life, politics included, to the point where even the "legitimacy and task of the state becomes bound exclusively to economic growth, global competitiveness, and maintenance of a strong credit rating," the alarming consequence of which witnesses the thinning of "liberal democratic justice concerns."[49]

Environmental harms are aggravated by a deeply compromised political process. The neoliberal demand for less government intervention is indeed a demand for more government manipulation of the market so as to skew the market mechanism in favor of powerful interest groups. In this regard the ideological work of neoliberal rhetoric comes from the way in which it effectively conceals this contradiction. In the same vein, inclusive green growth embraces government involvement in greening national economies, all the while ignoring how power shapes governance the world over. Predictably, under these circumstances green growth will never become inclusive. This is because the dominant political modality is depoliticizing.

At its core inclusive green growth is an economic outlook that presents an important alternative to high carbon emitting and polluting practices associated with business as usual, offering forth an environmentally friendly economy powered by renewable energy, clean and efficient technologies, natural-resource conservation, and green goods and services. However, the focus on using market mechanisms and the principle of economic efficiency to solve the problem of rising global greenhouse gas emissions, environmental degradation, and inequitable social relations overlooks the myriad ways in which power, privilege, and influence intersect across a variety of sociopolitical scales.

The prevailing obedience to a logic of competition and a political modality of neoliberal governance that demands nothing but endless servitude to capital accumulation is intensifying global inequity and environmental degradation. Rather than gently poking the bear of capital with an economic stick, choosing to engage with its competitive logic by using neoliberal governance as the primary strategy to overcome social and environmental problems, we need to poke the bear harder and with a bigger political stick. In other words, we need to acknowledge capital is its own animal with its own internal logic of competition that allows capital to function as capital. So while we need to continue engaging with the bear in order to diffuse the growing power and influence of elites over *who* governments speak for, two effects of which are environmental degradation and growing inequity, we cannot afford to underestimate the political depth of the problem as one involving *how* governments speak to the beat of a neoliberal drum. At the risk of sounding prophetic, social and environmental movements certainly have their work cut out for them. They need to both speak more loudly and more strategically, reimagining how politics works, creating and testing alternative political paradigms that are both practical and utopian, and inventing unusual political trajectories that move in ways that are as equally supple as neoliberal governance so as to successfully fend off appropriation by the very forces they are trying to surmount.

3

GREEN SCARE

What is an environmental emergency? Is it temporally finite, a break in the habitual rhythms of everyday life? How do you know when it hits? Does it suddenly suspend time, or does it speed it up? Or does the time of emergency linger to the point where it becomes inaudible and invisible, seamlessly blending in with routine rhythms and activities? Who responds to the urgent demands of emergency beginnings, and are these necessarily unprecedented? Or are they contingent beginnings, both continuous reminders of what currently exists coterminous with a discontinuous moment of rupture? How do the ripple effects of decisions made under critical circumstances produce new realities or, rather, and more ominously, expose grave secrets hunkering down deep in the dark crevices of normalcy?

It is now general knowledge that the scientific consensus is that carbon emissions arising from human activities are causing the climate to change. If allowed to continue, the Intergovernmental Panel on Climate Change predicts the earth's surface temperature will continue to rise over the course of the twenty-first century, explaining it is "very *likely* that heat waves will occur more often and last longer, and that extreme precipitation events will become more intense and frequent in many regions. The ocean will continue to acidify, and global mean sea level to rise."[1] Simply put, human activities are destabilizing the delicate regulation of planetary life by life for life. As James Lovelock has remarked, "man's industrial activities are fouling the nest and pose a threat to the total life of the planet which grows more ominous every year."[2] The result is the increased severity of natural hazards, resource depletion and scarcity, population pressures (growth and density in urban areas), declines in agricultural yields, and economic stagnation. When combined, these factors form what the United Nations has called the risk nexus of climate, water, energy and food security, which when taken together will increase the likelihood of forced migration (environmental, climate, and rural to urban migration), the loss of livelihood, and conflicts over resources.[3]

Unsurprisingly, in the past decades governments and international organizations such as the United Nations have increasingly declared a state of emergency in response to environmental events and conditions. In 2011 the worst drought since the 1950s parched the Horn of Africa, causing severe malnutrition that was linked to conflicts breaking out along the Ethiopian-Kenyan border and an increase in food prices. The United Nations has described the affected areas as a humanitarian emergency, reporting a "crude mortality rate of more than two deaths per 10,000 per day" and "58 percent of children under the age of five" to be "acutely malnourished."[4] In October 2013 the New South Wales premier Barry O'Farrell declared a state of emergency when more than fifty bushfires scorched their way through the Blue Mountains, destroying more than two hundred homes, burning wildlife, flora, and fauna to a crisp.[5] One month later, Australia's neighbors in the Philippines were hit by Super

Typhoon Haiyan, affecting in excess of 9.8 million people, killing 7,354 and destroying approximately half a million homes.[6] Philippine president Benigno Aquino III promptly declared a state of calamity. The following year, when severe rains fell in the Oso area of Washington State (United States), fifteen million cubic yards of soil, rock, clay, wood, and other debris slid down the mountain, engulfing the small town at the base, killing eight people. On March 22, 2014, Washington governor Brad Owen of Snohomish County declared the area under a state of emergency. Two days later, President Barack Obama signed an emergency declaration order authorizing the "Department of Homeland Security, Federal Emergency Management Agency (FEMA), to coordinate all disaster relief efforts" with the aim to save "lives and to protect property and public health and safety, and to lessen or avert the threat of a catastrophe."[7]

Nation-states and international organizations have been hard at work assessing the social, political, and economic implications of environmental security. The Swiss Environment and Conflicts Project has identified four stages to the "typical causal pathway to conflict."[8] The first is dependency on natural capital. The second is environmental scarcity. The third is environmental discrimination resulting from unequal access to natural resources. And last, "ecological marginalization when unequal resource access and population growth combine to drive further degradation of renewable resources."[9] Of deep political concern is how to create peaceful outcomes from environmental and human security threats. This usually entails a multifaceted and collaborative approach that engages local, national, and regional stakeholders, the military, and economic development initiatives. In an international geopolitical context this typically involves establishing transboundary initiatives and institutions.

One example would be the Fergana Valley project in central Asia, where the United Nations is working on fostering a program for the collective management of transboundary resources and the collection of data. Its aim is to build local and transboundary trust and cooperation.[10]

As it goes, the problem of environmental and human security is *"not whether people actually have a reason to commit violence, but what enables them to carry it out under particular circumstances."*[11] According to this framework, conflict is the result of *grievance* or *greed*, with the latter considered to be a more common source of discontent that leads to violence. This is because greed, which includes the greed for power, is the result of individuals using violence to extract economic profit and as a means of control. For instance, in arid parts of the world such as Syria and Iraq, water is a central weapon of war, with Sunni Islamist rebels, the Islamic State of Iraq and Syria, taking control of the Mosul dam and denying water to others, such as the Shia communities in the south of Iraq and Iraq's Kurdish and Christian communities.[12]

There are primarily three positions that respond to the question of emergency politics. The first is the state of exception theory. It maintains that sovereignty is established when law is suspended. Survival under these conditions is inherently undemocratic because, as the second position argues, the deliberative process is compromised if not shelved altogether. The third camp sets out to de-exceptionalize the exception, understanding the politically transformative potential of emergency affects and arguing that emergency states offer opportunities to democratize sovereign power. When considered together, as I do in the following, a useful theoretical apparatus with which to critically engage with a temporal politics of imagination emerges.

The first view, associated with the Nazi jurist Carl Schmitt, maintains that when a sovereign invokes their emergency powers, they violate neither the rule of law nor individual rights because both apply only under normal circumstances. The juristic thrust of Schmitt's position is that the essence of state sovereignty consists in having a "monopoly to decide" not the "monopoly to coerce or to rule."[13] The corollary can be summarized in this oft-cited maxim: "Sovereign is he who decides on the exception."[14] An emergency situation is an abnormal circumstance calling for extralegal measures that result in the inevitable suspension of quotidian rules and norms.[15] Bonnie Honig has called Schmitt a decisionist, because his position is that the "daily deliberative democratic undermining of friend/foe politics screeches to a halt, and a sharpened sovereignty is installed."[16] Schmitt's view is that a state of emergency permits governments to exercise exceptional powers, making the logic of exception the very meaning of sovereignty.

Still arguing that emergency politics results in a suspension of law and the institution of sovereign power are those who deconstruct the very dichotomy of normal versus exception, disclosing the continuity between them.[17] Naomi Klein has mixed the view that national crises produce undemocratic outcomes with psychoanalytic insights.[18] She charts the affective potential of emergency states under "disaster capitalism," describing how powerful elites use the confusion, disorder, and disorientation crises incite to institute otherwise unpopular neoliberal free market reforms. However, not all oppositional forces are entirely overcome; as such, the suspension of the law that reigns under the state of exception carries a political excess—that is, dissenters who simply disappear into indistinction, existing as bare life, to borrow a term used by Giorgio Agamben, both inside and outside the juridical order.[19]

Klein's perspective on the shock doctrine introduces a "killing machine" into social life, indicative of the state of exception, as Agamben explains. Using the juridical figure of Roman law, *homo sacer* (sacred man), Agamben describes a person who exists on the threshold between life and death (refugees or a person on death row), someone who can be neither executed nor sacrificed. In this way, the figure of the *homo sacer* illustrates the structure of the sovereign relation.[20] On the one hand, political life (*bios*) excludes bare life (*zoe*) from the juridical order, and on the other hand political life necessarily includes bare life. It is here in the sovereign's power to turn a subject of rights and citizenship into mere bare life (*homo sacer*), where totalitarian power relationships indicative of the concentration camp coincide with the exercise of sovereign power in Western democracies. One need only be reminded of the Bush administration collapsing the constitutional democratic framework of separate powers— executive, legislative, and judicial—which resulted in the indefinite detention of those whose activities were deemed a threat to national security. Under the USA Patriot Act of 2002 people could be denied a juridical identity. That is, they were indefinitely detained neither as a prisoner of war nor under arrest. Under these conditions environmental activism became synonymous with terrorism. It is the perfect example of how "law and legal reasoning not only give form to the economic," as Foucault maintained, "but economize new spheres and practices," as Wendy Brown has argued.[21] As Brown has formulated it, "law becomes a medium for disseminating neoliberal rationality beyond the economy, including to constitutive elements of democratic life. More than simply securing the rights of capital and structuring competition, neoliberal juridical reason recasts political rights, citizenship, and the field of democracy itself in an economic register; in doing so, it disintegrates the very idea of the demos. Legal reasoning thus complements governance practices as a means by which democratic political life and imaginaries are undone."[22] The FBI classification of ecoterrorism expands the definition of terrorism provided by the United Nations as crimes and violent acts against civilians to also include the targeting of property with the goal of being intimidating or coercive: "The FBI defines eco-terrorism as the use or threatened use of violence of a criminal nature against innocent victims or property by an environmentally-oriented, subnational group for environmental-political reasons, or aimed at an audience beyond the target, often of a symbolic nature."[23] In essence, the situation of the ecoterrorist captures the struggle of politics proper in all its divisiveness and antagonism as it wrestles with the ideological structure of social reality as a never-ending state of emergency. To recall Slavoj Žižek here, "Ideology is not simply a 'false consciousness,' an illusory representation of reality, it is rather this reality itself which is already to be conceived of as 'ideological.'"[24]

As the climate continues to change low-lying island nations will disappear off the map, food shortages and the cost of food will increase, potable water supplies will grow scarce, and generally the world will become a more inhospitable habitat for humans and most other species, not to forget the willful destruction of human habitats as a means of waging war, as I discuss in chapter 7. Obviously, the poor, who have fewer resources, are especially at risk. This is a violent state of affairs. People who have nothing to do with causing the problem of climate change (future generations, islanders, the poor, who barely leave a carbon footprint at all) will be some of the worst affected.

This violent state of affairs has led to the formation of numerous eco-activist groups (Greenpeace, Earth First, 350.org, and

Rising Tide, to name a few). All are working hard at organizing, strategizing, and interrupting the supply and distribution
chain of carbon-intensive industries and products, as well as
protesting against the unregulated and unabated increase in
global greenhouse gas emissions and its effects. And yet a curious phenomenon is occurring with the emergence of ecoterrorism on the national security radar: the criminalization of
environmentalism. Rather than governments instituting concrete
solutions that will abate the environmental emergency along
with all the social and economic vulnerabilities this entails,
world leaders are instead exercising their emergency powers to
restrain, infiltrate, and place under surveillance the activities
of environmentalists who are working hard and peacefully to
draw attention to the violence of climate change and environmental degradation.

One example that comes to mind is the case of Mark Kennedy (Mark Stone). Stone began his career as an undercover
police officer spying on environmental activists in 2003 when he
joined the anarchist group Earth First, later working with Dissent and the G8 Summit protests in Gleneagles. By 2006 he had
become an integral player in the organization of transport for the
Camps for Climate Action. Working for the United Kingdom's
National Public Order Intelligence Unit, his assignment
was simple: infiltrate and monitor "domestic extremists."[25] He
posed as a climate change activist for seven years, and his work
led to the arrest and conviction of twenty-nine activists who
interrupted the train delivery of a thousand tonnes of coal to
Drax, and in 2009 of twenty-six environmental campaigners the
night before they planned on occupying the Nottinghamshire
coal-fired Ratcliffe-on-Soar power plant. Stone's activities as
an environmental spy directly undermined the political edge of

eco-activism, reinstating the ideological structure of emergency politics.

The political force of eco-activism comes from demonstrating how everyday life is inexorably ideological. In this light, the naturalization of the emergencies environmental degradation presents is in itself an ideological phenomenon. The strength of environmental activism comes from a refusal to accept ecological deterioration as just another instance of anarchical nature. Rather, the politics of environmentalism arises from denaturalizing the environmental challenge by exposing its ideological structure.

As an illustration, on March 4, 2014, Greenpeace activists hung banners at the Procter and Gamble headquarters in Cincinnati that read, "Head and Shoulders wipes out dandruff and rainforests" and "Head and Shoulders: stop putting tiger survival on the line." The banners brought home a clear message: palm oil is a key ingredient in the company's antidandruff shampoo (as well as in Olay skin care and Gillette shaving gel products). The growth of the palm oil industry is leading to the clearing of rain forests. This is connected to climate change and the extinction of species such as the Sumatran tiger and orangutans.[26] All nine activists were charged with burglary, vandalism, inducing panic, and trespassing. The felony charges brought with them up to nine and a half years in prison, an excessive sentence for what they did. It was not until December 2014, after Procter and Gamble pressured the prosecutor to drop the charges, that their legal situation changed: a lesser misdemeanor charge of criminal trespassing (eighty hours of community service and one year probation). Until December pressure from the activist community, myself included, had fallen on deaf ears. It took a corporate power to turn the tide, and the irony is that

Procter and Gamble emerged the hero and the activists the criminals.

The Greenpeace protest case exposed the violence of immanent legal threat and the paradoxical relationship this holds with peaceful protest being protected under the First Amendment of the U.S. Constitution and the political muscle of the corporate sector. Understood as what Žižek terms an obscene supplement to the scaffolding of power, the Greenpeace activists constituted a political subject as the effect of power's antagonistic excess.[27] As the supplement of power, the political act consciously brought the unnameable, unintelligibility of the real into relation with the symbolic dimension of political power. In this regard the Greenpeace activists exposed the emergency state along the frontier between the real and symbolic. The political force of the act came from how the activists agitated that frontier. The truth of the environmental emergency is thereby exposed not as a representation, one that translates the situation into a security issue, or the spectacular images of devastation and death flooding the media, or even the alarming statistics reported by international and nongovernmental agencies. Truth is revealed where there is no representation, from where representations are produced—in other words, in this instance where the state of exception is legitimated.[28]

The second position argues in favor of devising political frameworks that enhance deliberation, a central feature of liberal democracy. The goal of the deliberative position is to uphold the rule of law despite the urgency and disorder emergencies invoke—for example, the problematic at the core of Michael Ignatieff's ethical interrogation of how the United States responded to the threat of terrorism after 9/11.[29] Ignatieff insists it is morally questionable to surrender a prior democratic commitment to protect individual rights and the rule of law in the name of national security. Regardless of how justified this may seem at the time, it always creates a moral remainder. Ignatieff mounts a compelling case in favor of using a former democratic guarantee of due process for reviewing governmental decisions made during a state of emergency, and to democratically oversee measures taken in response to an emergency.

Like Ignatieff, Elaine Scarry has tried to reinvigorate the mechanisms of a liberal democracy as a way to deal with emergency states. The expression "under a state of emergency" has been used to alert and alarm populations, effacing the deliberations of democratic governance, Scarry explains. She states, "In an emergency, the habits of ordinary life may fall away, but other habits come into play, and determine whether the action performed is fatal or benign."[30] She questions the common belief that social and political structures are lost during an emergency and the related idea that views everyday life and emergencies as polar opposites, whereby the former is habitual and the latter nonhabitual. The three dominant views of emergency Scarry debunks are (1) action does not involve thinking; (2) to act quickly means there is no time to waste thinking; and (3) thinking involves setting habits to one side. The first two, she explains, presuppose thinking is abandoned in the name of action and, third, that habit is abandoned in the name of thinking. However, habit is apparent in emergency preparations just as much as thinking involves habit. "The question is not *whether* habit will surface in an emergency (it surely will)," she writes, "but instead *which* habit will emerge, and whether it will be serviceable or unserviceable."[31] Mounting a convincing case in favor of preparedness training, she explains that "far from being structureless, a crisis is an event in which structures inevitably take over." Will they be positive or negative structures, she asks. And will they ensure a "just distribution of authority and risk"?[32] Moreover, will they ensure a commitment to all surviving equally?[33] Ultimately, deliberation is key to ensuring that good constitutional procedures are instituted and that the maintenance of important democratic constitutional arrangements is upheld.

However, this position assumes there is a transparent and legitimate process of deliberation currently driving democracies around the world. The reality is quite different.[34] Since 9/11 intelligence gathering has been increasingly outsourced to private agencies, with information being readily shared with the private sector. As sovereign powers ramp up their surveillance of environmental groups, so too does the corporate sector. There are a growing number of allegations from environmental organizations and investigative news outlets describing the manner in which the private sector is now overextending its powers, infiltrating and disrupting the coherency of environmental groups with a view to strengthening and shielding the corporate sector from public scrutiny and accountability.

As the corporate sector infiltrates environmentalist groups, as much as states do, a disturbing political nexus of capitalism, liberal democracy, and violence appears. To borrow from Retort's analysis of the Iraq War, it is a "radical, punitive, extra-economic restructuring of the conditions necessary for expanded profitability."[35] Clearly the difference between elected governments and the corporate sector is blurry. The upshot is that environmentalists now undergo a double act of surveillance—by governmental agencies, such as the Pentagon in the United States and MI5 in the United Kingdom, as well as by undercover transnational corporate intelligence gathering—for example, the three UK energy companies (E.ON, Scottish Resources Group, and Scottish Power) found hiring private security firms to spy on environmentalists.[36] Surveillance as a corporate strategy completely undermines the framework of deliberative democracy

Scarry attempts to safeguard because it means that the ground upon which civil society deliberates political issues is already impaired.

Last, and the most promising position for environmental activism, is the view of de-exceptionalizing the exception. The idea here is to combat the ennui of politics as usual by invoking the affective and populist underpinnings of sovereign legitimacy. For Honig this involves "democratizing emergency" by "seeking sovereignty, not just challenging it, and insisting that sovereignty is not just a trait of executive power that must be chastened but also potentially a trait of popular power as well, one to be generated and mobilized."[37] Here Adam Bandt provides a theoretical compass for Honig. He argues for the "emancipatory potential of emergency."[38] If the "flesh and blood battles" of creative labor are let loose, he explains, emergency states could provide a real opportunity to generate new collective modes of living that might move us past mere survival mode.[39] To do this we need to be willing to reclaim the affective and liberatory potential emergencies offer us. Their advice is simple: use emergencies as a transformative constituent power.[40]

The political pronouncement of a "state of emergency" occludes the fact that emergency is the engine of liberal democracies, which are in an endless emergency state arising from the all-inclusive power structure of global capitalism that includes the source of its own transgression in the work of the environmental activist. When the black hole of the real of climate change and environmental degradation irrupts, the excesses of liberal democracies are confronted head-on. In response world leaders anxiously institute a state of emergency in an effort to conceal a situation that ultimately lies beyond the rule of law. When leaders of the "free world" refuse to confront the challenge of the real disrupting the order of everyday life, the "presymbolic substance" of the social, economic, political, and cultural spheres in all their "abhorrent vitality" (to unabashedly lift from and resituate Žižek's discussion of the Lacanian real) that environmental disasters present remain unexposed to political critique.[41] Utterances of "emergency" and "public security" and "exceptional times call for exceptional measures" work to unify the social field around, albeit an empty, master signifier of security and resiliency, depoliticizing the emergency at hand. These are ideological declarations that create political effects such as increased surveillance and the militarization of day-to-day life.

If what we mean by politics is an experimental combination of ideas and acts that transform and open up an otherwise closed and rigid social field, then all that the discourse and practices of environmental emergencies and security stir forth is a new kind of fascism—environmental fascism; that is, if what we mean by fascism is, as Gilles Deleuze and Félix Guattari have described it, "There is fascism when a *war machine* is installed in each hole, in every niche."[42] Continuing on, they present the following unnerving diagnosis: "What makes fascism dangerous is its molecular or micropolitical power, for it is a mass movement: a cancerous body rather than a totalitarian organism . . . Only microfascism provides an answer to the global question: Why does desire desire its own repression, how can it desire its own repression? The masses certainly do not passively submit to power; nor do they 'want' to be repressed, in a kind of masochistic hysteria; nor are they tricked by an ideological lure . . . It's too easy to be antifascist on the molar level, and not even see the fascist inside you, the fascist you yourself sustain and nourish and cherish with molecules both personal and collective."[43] I return to this problem of the masses desiring their own oppression in the following chapter. For now I am interested in examining how the declaration of a state of emergency in response to natural disasters implies that these are exceptional events, when in fact climate change and the uninterrupted march of global capitalism make them everyday events.

We fetishize an everyday occurrence when we make it an exceptional state of "emergency," for we invest our creative political energies and transformative collective affects such that the subject of emergency is no more real than the libidinal economy driving it. As Žižek points out in his reading of Jacques Lacan, "what the fantasy stages is not a scene in which our desire is fulfilled, fully satisfied, but on the contrary, a scene that realizes, stages the desire as such."[44] Continuing on, he clarifies, "The realization of desire does not consist in its being 'fulfilled,' 'fully satisfied,' it coincides rather with the reproduction of desire as such."[45] This brings us to the kernel of environmental fascism: anxiety. Žižek writes, "Anxiety occurs not when the object-cause of desire is lacking, it is not the lack of the object that gives rise to anxiety but, on the contrary, the danger of our getting too close to the object and thus losing the lack itself. Anxiety is brought on by the disappearance of desire."[46]

The question framing environmental security policy is, where do environmental threats lie? As the violence of rogue states or militant groups spectacularly flood the media and take up center stage in geopolitical policy and military strategizing, the "objective violence" of global capitalism, as Žižek has called it, is let off the hook.[47] There is an excessive nucleus to global capitalism that comes in the form of impoverishment, ecological devastation, social instability, death, and toxicity. Walter Benjamin's remark that the "tradition of the oppressed teaches us that the 'state of emergency' in which we live is not the exception but the rule" can be applied to the contemporary condition of ecological deterritorialization.[48]

What is at stake with environmental extremities? How are they encountered, as object or affect? The pressing problem all life on earth is facing with climate change is less spatial than it is temporal. In other words, there are numerous temporalities to the emergencies posed by environmental degradation. First, there are short-term extremities, such as those that arise when natural disaster hits. Then, there is the nonteleological and prolonged steady state of ecological forces activating multiple affects, forming new connections with energetic systems (waves, El Niño, melting ice caps) and destabilizing the fragile equilibrium balancing life on earth, pushing it to the tipping point. These work in conjunction with the forces of globalization, capital accumulation, and militarism conjugating and provisionally stabilizing one another. What do we do with the new relations, affects, and energies that emergency states invoke? Viewed this way, codifying the openness of these states as a security issue basically militarizes the situation. It may restore order to an otherwise anarchical situation, returning continuity to the violent ruptures global capitalism is inflicting, but it also forecloses the situation off to its own immanent creative potential.

How can environmental activism possibly be effective with displacement activities of this magnitude? The three responses to the question of emergency politics previously described provide some guidance on this problem, but each shares the same conceptual deadlock. They all assume a conceptual boundary between the sovereign and the citizenry. All presuppose in one way or another that sovereignty represents the citizenry either through the Hobbesian choice of the people giving up some of their freedoms in the state of nature to establish a sovereign who will ensure their security, or through an inclusive process of civic deliberation that will ensure the citizenry is adequately represented, or as a representative of the creative labor of the citizenry to transform their political realities and democratize the figure of the sovereign.

Yet the political force of sovereignty is not just a matter of representation, the effect of which would be to relativize politics and strip it of its materialist dimension. Sovereignty is not just an entity, an office, or political figure, it is a process that is affective, and it is the effect of a particular investment of social energies and forces, or what Deleuze and Guattari called desire.[49] As a process sovereignty can be both active and reactive. This raises a problem of political origins. Does sovereign power arise from the political process (deliberative and creative labor) or from itself (state of exception)? The dichotomy here is in itself problematic for the sovereign represents the people and also institutes a people, yet framed this way the relationship between sovereignty and the people still doesn't allow us enough scope to "revitalize the very meaning of the political out of the torment of its catastrophic condition," as Evans and Reid have justifiably demanded.[50]

For Evans and Reid, "what is at stake here is not simply the 'aesthetics of existence' wherein life conforms to some glorious representational standard of beautification . . . What is demanded is the formulation of alternative modes of existence that are not afraid to have reasons to believe in this world" and, I would add, reasons not to believe in the end of the world.[51] If we approach this imaginatively—that is, as both an idealistic and realistic exercise—we take a leap of faith in the abstract political force of sovereignty as a re-presentation of the liberatory energies of the multitude, and we ground this in concrete material affects.[52] Sovereignty is therefore both determined as a result of political engagement and a determinant of it. It is both inside and outside the political process. It is both a pregiven political force and a force that invests the desires of the social field in an articulation of the political. Political engagement thus arises through the asymmetrical relationship of sovereignty and a multitude, experimentally bringing the abstract realm of political concepts, ideas, and thoughts into relation with the concrete materialities of political acts. When sovereignty and a multitude combine, a new open political imaginary that is as much idealist as it is realist can materialize, an issue I examine more closely in chapters 8 and 9. Until then, the basic lesson drawn from emergency politics is that environmental activism remains an important political locus; it is just that it is yet to find its proper movement in the time of transformative politics.

4

FASCIST EARTH

*Ain't no power like the power of the people, cause the power of
the people don't stop!*
  *Ain't no power like the power of the sun, cause the power of
the sun don't stop!*

—Climate March chants

hen we speak of the environmental movement,
what exactly are we referring to? Globally speaking, there are thousands of environmental coalitions, networks, organizations, funds, forums, associations,
societies, agencies, institutes, counsels, and blogs. There are
millions of small community programs, committed individuals who identify as greenies, and ad hoc environmentalists
who don't belong to any specific environmental organization
but whose conscience draws them out onto the streets to protest any number of environmental issues. In short, the environmental movement is a multidimensional and amorphous
political phenomenon.

The "movement" consists of a variety of messages and issues
that arise across different geographic locations—climate change,

logging, antiwhaling, fracking, water rights and safety, GMOs, toxic pollution, dirty energy, ocean health, sustainable fishing, nuclear energy, and species extinction. Therefore, although we refer to "the environmental movement," the movement is far from being a unified entity. There are movements within the movement, each of which is distinct yet related. Singular movements within the overall movement referred to as the environmental movement change as they connect. The movement also transforms as the number of participants, leadership, and its sociopolitical context change. Yet despite a differentiating and self-organizing character, the backbone of the movement remains singular: stopping environmental degradation.

As a political practice, environmentalism infuses new energy into an otherwise "anemic *homo politicus*" that, as Wendy Brown has formulated, has been reconfigured into *homo oeconomicus* under neoliberal rationality,[1] and which Kenneth Surin has portrayed as the neoliberal constitution of subjectivity.[2] Under such circumstances, this chapter argues, one of the greatest hopes for democratic politics lies with the environmental movement. If what we understand by environmentalism is a variety of sociopolitical practices seeking more inclusive, equitable, and caring societies where the concept of sociality is expanded to include a variety of species, the physical conditions that sustain them, and the relationships between them, then environmental politics is well positioned to change how exploitation and oppression are normalized.

However, social conditions also produce particular kinds of political thinking and practices that are not necessarily emancipatory. In chapters 1 and 2, I investigated how neoliberal rationality and governance co-opt and displace the emancipatory promise of environmentalism. In this chapter, I examine emancipatory and reactionary forms of environmental thinking and practice. In formulating this argument, I present a distinction between environmental politics and environmentalism in politics. The latter is neither intrinsically emancipatory nor inclusive. The former is. In this chapter I follow the convergence of right-wing extremism and environmentalism. This is not a recent phenomenon. It has a long history that extends back through Nazi ideology to romantic notions of a *Volk* and even into ancient times to Hippocrates.

In light of this, environmentalism needs to be quick on its toes, alert to compromise from both within and outside the movement. This does not involve a clear choice between grassroots and top-down initiatives. Rather, it entails engaging in bastard solidarities that use different political strategies. Instead of choosing between reformist and revolutionary political practices and ideas, the emancipatory promise of environmental politics might be strengthened by engaging both political strategies, not to be confused with working both sides of the political spectrum. Indeed, as I argue, working with and adopting ideas from the far right does not strengthen, rather it displaces the emancipatory promise of environmentalism.

ENVIRONMENTAL POLITICS

As outlined in chapter 1, neoliberalism has resulted in extreme inequities. It has also led to a global financial meltdown prompted by the U.S. subprime mortgage emergency, as I discuss further in chapter 5. Furthermore, as outlined in chapter 1 the environmental harms that widespread consumption and endless economic growth have produced are now being turned into another free market opportunity where a price is placed on negative environmental externalities. This situation is the result of two forces working in tandem. The first, as discussed in chapter 1, is economic, and it is the result of endless capital accumulation. The second, as outlined in chapter 2, is political, what Foucault depicted as liberal governmentality. "Today's regime of capitalist accumulation and the neoliberal and neoconservative ideologies identified with its current ascendancy," Surin has stated, "simply have no need for the classical Citizen Subject."[3] In his estimation, the scale and scope of corporate power have expunged the political subject of the liberal democratic state.

Invoking the concept of neoliberal rationality, Brown has examined the corrosive effect it has had on democratic life. Briefly stated, Brown concludes that the "most important casualty of the ascendance of neoliberal reason" is *homo politicus*.[4] As the definition of a person has been increasingly tied to a concept of human capital, the subject's relationship to both itself and the state has been reoriented: "Rather than a creature of power and interest, the self becomes capital to be invested in, enhanced according to specified criteria and norms as well as available inputs . . . No longer are citizens *most importantly* constituent elements of sovereignty, members of publics, or even bearers of rights. Rather as human capital, they may contribute to or be a drag on economic growth; they may be invested in or divested from depending on their potential for GDP enhancement."[5] Interpreting and translating everyday social, cultural, political, economic, and personal life using an economic frame diminishes the vitality of the demos, because what this in effect does is supplant *homo politicus* with *homo oeconomicus*.

Writing several decades earlier, social ecologist Murray Bookchin identified a similar problem to the one Brown and

Surin articulate, but he gave it an ecological twist. The "real bat-tleground on which the ecological future of the planet will be decided," he wrote, "is clearly a social one."[6] "In effect," he con-tinued, "the way human beings deal with each other as social beings is crucial to addressing the ecological crisis."[7] More specifically, for him the ecological problem was symptomatic of a competitive market-driven society, unbridled corporate power, and the subsequent hierarchical organization of the social field. Considered in tandem, this situation has gradually disempowered people to effect political change.

Bookchin blamed the environmental crisis on a social con-text that offered no political alternatives. He maintained that ecological problems require social solutions, and in order for this to happen democratic political life needs to be radically recon-figured.[8] His solution: harness the creative power of human beings to establish a cooperative society of libertarian munici-palism, where local civic power replaces centralized state power. In short, localize democratic politics.

The political alternative to a state-based party system of elec-toral campaigning, Bookchin offered, is a networked political structure consisting of free council confederations, locally based elections, and public assemblies. In his writings, he presents a communitarian vision of a cooperative society that extends to all facets of life, including greater complementarity between humans and nature. He basically reinterprets the democratic form, inverting the top-down party system of bureaucratic state-craft into a bottom-up arrangement of direct democracy based on local popular assemblies.[9]

A good example of what Bookchin envisaged would be the nineteen-year-long Women's Peace Camp at Greenham Common. On September 5, 1981, a group of women and their children arrived at Greenham Common, England, in protest against the housing of ninety-six American nuclear cruise missiles at the Royal Air Force base at Greenham. What began as a lone protest march quickly turned into an open-ended campaign as four women, chosen by the group to represent different generations of women, chained themselves to the fence at the base.[10] The group set up camp outside the base. Over time numerous sites were established, remaining until 1990 when the missiles were removed, with the last of the peace campers leaving on September 5, 2000, after it was confirmed a memorial commemorating the Women's Peace Camp would be built at the site.[11]

Over the years the Women's Peace Camp grew, including women from all walks of life, across generations, and from around the world. David Fairhall, reporter on the Cold War and the *Guardian*'s defense correspondent, has described how the movement transformed the lives of many of the women who partook in the demonstration. With a tone that is troublingly moralizing, one that exposes a deeply masculinist perspective, he characterized one of the early protagonists of Greenham Common, Helen John, as abandoning her family. She fell "in love with a woman," he writes in his report, and asked "for a divorce," exchanging "domestic respectability for the uncomfortable life of a political activist."[12]

Fairhall's description obfuscates how the women of Greenham Common used the image of respectable domesticity to their own political advantage. Photographs taken during that time show women knitting, singing, mothering, conducting sit-ins, fence chainings, forming human chains, confronting police and the military, cooking, camping out, and forming human blockades. Living at the camp as a self-organizing female community was a political strategy that combined domesticity with activism, thus subverting the dominant view of domesticity as private and passive.

Grandmothers, mothers, wives, lesbians, single women, children, teenagers, celebrities (Yoko Ono) all participated in the protests with as many as thirty-five thousand reportedly linking arms around the site in December 1982 to stop preparations for the base.[13] And even more came together in April 1983, with the *Guardian* reporting seventy thousand women from the camp had formed a human chain extending fourteen miles and encompassing three nuclear weapons centers.[14] The women of Greenham Common withstood evictions by police, arrest, verbal abuse, rough living conditions, and adverse weather. Throughout all this, they firmly adhered to a principle of nonviolence.

From the very beginning the campaign involved mothers concerned about the future of their children in a world where nuclear armaments had become the norm. They made a simple decision: stop nuclear proliferation. One image "screened around the world at political meetings and consciousness-raising groups" is a snapshot of a human chain.[15] It depicts a woman linking hands with a toddler, a young child, and another woman. In this way, domesticity and the role of motherhood were used as political tools to challenge the machismo of militarism. At home in the fields where they camped, children were cared for, everyone looked after everyone else, and even babies were born under the care of midwives. A new economic order emerged as people from outside the camp helped feed and meet the needs of the women inside the camp. The women called public attention to a domesticity that no longer limited political autonomy, turning the traditional role of caregiver into a political opportunity: protect all life on earth.

Greenham Common became a site where a variety of political orientations mingled feminist, environmental, peace, and antinuclear movements. The environmental politics of the women at Greenham Common directly challenged numerous structures of exploitation and oppression—gender relations, sexuality, the environment, generational power, patriarchy, competition, and militarism. Staunchly leaderless, the women collectively cared for one another, made decisions together, and worked hard at ensuring a broad, inclusive public face to the movement. They refused to mobilize a hierarchical governance structure. There was no single woman who "represented" the group. They even rebuffed the homogenizing stereotype of "Greenham woman" for failing to represent the "thousands of women who came through the camp, or supported it as best they could from a distance."[16] Together they disrupted the democratic frame of nuclear deterrence dominant at the time, breathing new life into the democratic notion of people ruling themselves as an organized community.[17]

Environmentalism has a long track record of challenging institutional truths and destabilizing the power relations that secure them. Some of the most charged examples of environmental politics coalesce around instances when "bodies show up or move through" space "in ways that are not allowed, or when communities form on either side" of a divide, to paraphrase Judith Butler and Athena Athanasiou's discussion of the logic of dispossession and resistance to it.[18] More recently, environmental groups have aligned with the Black Lives Matter movement in the United States and with humanitarian organizations assisting the Syrian refugees.

Greenpeace teamed up with Médecins Sans Frontières to assist Syrian refugees who risked their lives crossing the sea between Turkey and Greece. From November 28, 2015, to March 23, 2016, the two organizations helped 18,117 people and conducted 361 operations off the Greek island of Lesbos.[19]

After of slew of killings of young black Americans at the hands of police, leading figures in the American environmental movement quickly responded to offer their unequivocal support for Black Lives Matter. Michael Brune, executive director of the Sierra Club, issued the following statement: "The Sierra Club believes all people deserve a healthy planet with clean air and water, a stable climate and safe communities. That means all people deserve equal protection under the law and the right to a life free of discrimination, hatred and violence. Unfortunately those aspirations and goals are not a reality in our country, and that is why the Sierra Club stands in solidarity with all of those saying Black Lives Matter, demanding justice, accountability, and action to confront the racism and inequality that has allowed these tragedies to persist."[20] The next day, the president of the Environmental Defense Fund, Fred Krupp, made the following announcement:

> Environmental Defense Fund rarely speaks out on issues not directly related to our work. But since our work is dedicated to building a world in which people and nature can thrive together, we cannot stay silent in the face of the injustice that is robbing so many Americans of the chance to thrive.
>
> We deplore the police killings of innocent black people in Baton Rouge, Minneapolis and elsewhere, and the killings of law enforcement officers gunned down while protecting peaceful protesters in Dallas—in other words, while doing precisely what police officers are supposed to do.
>
> We believe that black lives matter, and we join the urgent and rising call for justice.[21]

Expanded participation, sharing of resources, recognizing common interests, and fostering collective understanding strengthen political capacity and effectiveness across a variety of political scales (local, regional, national, and international advocacy) and platforms (protests, education, legislation, policy, public and private programs, international treaties, and nongovernmental organizations). As such, the alliances formed with groups whose mission is not specifically "environmental" are a welcome addition to the practice of politics in the public sphere. Indeed, they indicate a democratization of the public sphere.

Forming alliances with other sociopolitical groups can potentially deepen and enrich the political principles and activities of all involved. Infusing the environmental movement with multiple discourses from other social movements brings fresh perspectives on environmental issues. Along with different interpretations and viewpoints on environmental affairs comes conflict. John Dryzek has explained that "different sides interpret the issues at hand in very different ways . . . the way the issue is dealt with depends largely (though not completely) on the balance of these competing perspectives."[22] Focusing on the discourse of environmental politics, he points out that environmental politics includes those who see themselves as environmentalists and also those who are "hostile to environmentalism."[23] To add to Dryzek, the way environmental issues are sometimes politically deployed triggers an especially hostile reworking of environmentalism.

If we glance in the rearview mirror for a moment, we see not so much a coherent linear political movement unified around an emancipatory environmental politics but, instead, many movements that cover the whole political spectrum, including the far right. At times the spirit of radicality has fueled ecofascists as much as it has left-leaning greenies. On occasion, throughout the not so distant past unholy alliances have formed with nativists. These are inconsistent with the emancipatory impulse of environmental politics.

ENVIRONMENTALISM IN POLITICS

Caution needs to be heeded over how environmental ideas are used. The dramatic localization of the political sphere in response to the twofold problem of ecological destruction and social unease is not inherently emancipatory. Indeed, right-wing extremists have appropriated and distorted Bookchin and other environmental thinkers in justification of an especially pernicious, segregationist, and racist ecopolitical program.

Part 7 of the National-Anarchist Manifesto begins by citing the work of Bookchin.[24] The manifesto lays out a clear environmentalist agenda as a way to solve the problems of society. Pushing the communal, natural, and collectivist ideas of environmentalism to the extreme, the manifesto proposes a return to the land: "In the past, man had an inextricable bond with the soil. Not only was his racial heritage of great importance, but he also knew how essential it was to carve out and defend a territory in which to express his own values and aspirations. Sadly, however, due to the immense destruction that has been wrought on the environment today, not least in the overpopulated countries of Europe and North America, it is impossible to live in harmony with nature without moving away from the cities and out into the countryside."[25] In what sounds like a direct appropriation of the ideas of Bookchin, National-Anarchists also envisage a decentralized society made up of communities that form a confederation as a solution to the parliamentary establishment. However, the cooperative, inclusive, and differentiating core of Bookchin's vision is distorted into a system of regional cooperation between "like-minded communities" whose shared values individuals "highly trained in the methods of self-defense and, if necessary, warfare" protect, and the arms they use "will be held in the hands of the community itself."[26] The notion of diversity is used against its fundamental principle of inclusivity. "Racial separatism," they state, "is the only way that the organic balance can be restored."[27] The principle of diversity is twisted into grounds for justifying ethnic and racial exclusion. "National-Anarchists must form new communities based on their own racial and cultural values. The maxim of the future will be respect for others and unity in diversity."[28]

At first glance the National-Anarchists appear to be a somewhat marginal radical organization, ineffectual and inconsequential in the larger realm of environmentalism. Yet the political activities of the movement are not limited to the online world of chat rooms, YouTube videos, and the like. National-Anarchists have participated in antiglobalization, pro-Tibet, pro-Palestinian, ecological, and animal-liberation campaigns across the globe, infiltrating left-wing demonstrations from Australia to the United Kingdom, the United States, and Europe. On September 8, 2007, they joined the anti–Asia-Pacific Economic Cooperation summit demonstration in Sydney. Enraged by an attempted right-wing infiltration of an otherwise left-wing event, left-leaning antiglobalization activists expelled the black bloc from the protest.[29]

The main ideologue for the National-Anarchist cluster is longtime ultraconservative racial separatist Troy Southgate. In an interview in 2012 Southgate described himself as a "decentralist," of neither left- nor right-wing political orientation,

an opponent of "State-orchestrated policies," and a racial separatist.[30] Southgate operated on a political platform of commu-
nity-based environmental activism combined with a fervently
nationalist and anti-immigration political program.

Some background on Southgate is in order. When he formed
the patriotic socialist party English Nationalist Movement
(ENM) in 1992, he set out to reinstate the fascist connection
between blood and soil. Graham Macklin has explained how
Southgate propounded a deeply right-wing view of native Anglo-
Saxon *völkisch* identity, holding up English pastoral life as an
ideal to aspire toward. At this time his goal was to "defend indigenous white culture from the 'death' of multiracial society."[31]

In 1998 as his politics underwent a conversion to anarchism,
Southgate disbanded the ENM, establishing in its place the
National Revolutionary Faction (NRF).[32] In his words, "The
NRF was a hardline revolutionary organization based on an
underground cell-structure similar to that used by both the
Islamic Resistance Movement (Hamas) and the IRA" aimed
at countering what he saw as "increasing violence against the
indigenous white community" at the hands of Asian immi-
grants.[33] His version of national anarchism was far removed from
"anarchism's humanistic social philosophy," which he "rejected as
'infected' with feminism, homosexuality, and Marxism."[34]

Southgate was convinced that Richard Hunt's vision of a
rural society consisting of small communes founded on family
values, united by the laws of kinship, and "protected by armed
militias" was the answer to the ills of contemporary life.[35] A pan-
anarchist, pan-secessionist, and participant of NRF, Hunt was
the founding editor of *Green Anarchist*. The editorial board
eventually removed him because of his reactionary views. Consequently, he went on to serve as the editor of the right-wing
magazine *Alternative Green* until his death in 2012. He helped

stage the Brighton Anarchist Heretics Fair in May 2000 that aspired to move beyond the dialectic of left and right, communism and capitalism, in what is referred to as the Third Position. The fair consisted of a right-wing alliance between *Alternative Green*, the NRF, the Wessex Regionalists, Christian anarchists of Albion Awake, medievalists such as Oriflamme, and the Anarchic Movement.[36]

It is important to recognize that Southgate's ecofascist leanings occur not only at the level of content—filling the democratic form with ecofascist content such as a return to a *völkisch* view of the world that draws on the tenets of environmental fascism under the German National Socialists—but also at the level of form—how the democratic form is put to work to channel fascistic energies and structures of interpretation. The problem is as much hermeneutic as it is machinic. It concerns both interpretation and how interpretation works.

The point here is not that Bookchin is an ecofascist; rather, environmental issues and ideas, as they are used in politics, are not inherently emancipatory. The political character of environmentalism all depends on how environmentalism is put to work. It is one thing to identify the source of socioenvironmental degradation with capitalism and the state and another entirely to turn an otherwise inequitable structural problem of exploitation and competition into a problem of immigration and racial contamination, the solution being a nostalgic return to the past, racial segregation, and a world filled with independent armed militias defending loyal converts from the evil threat of outsiders.

Other nativists have infiltrated even the most mainstream of environmental organizations. The Sierra Club at one point in its history endorsed the anti-immigration views of nativists, publishing *The Population Bomb* by Paul Ehrlich in 1968. Ehrlich linked the environmental problem with overpopulation, warning that if fertility rates in the underdeveloped world were not addressed, the environmental problem would worsen.[37] Invited to speak at the first Earth Day celebration on April 22, 1970, Ehrlich shared his apocalyptic view of an overpopulated world. From 1971 to 1975 nativist and sympathizer of strict population control John Tanton served as the chairman of the Sierra Club, bringing the club within a hair's breadth of turning into a right-wing organization. The Southern Poverty Law Center has reported the "extensive and hard-fought battle for control of the Sierra Club" by nativists that "would continue right through to 2005, when the last nativist attempt was beaten back decisively."[38]

There are numerous other examples. For instance, previously known as the Environment Fund, the Population-Environment Balance organization pushed for a "moratorium on immigration" in the United States.[39] In an attempt to woo over environmentalists, in 2008 the nativist organization America's Leadership Team for Long Range Population-Immigration-Resource Planning ran a series of full-page ads that appeared in "relatively liberal publications, including *The Nation*, *Harper's Magazine*, and *The New York Times*."[40] Betsy Hartmann has described this phenomenon as an instance of the "greening of hate."[41]

Yet the connection formed between right-wing thinking and environmentalism is not just a political anomaly. Racist nationalism has a long-standing history in ecological thinking. Frank Egerton has presented early evidence of the race-nation-environment correlation occurring as early as the fifth and fourth centuries B.C.E. in the Hippocratic Corpus. In paragraphs 12–20, "Airs, Waters, and Places," "environmental factors" are said to "determine racial conditions."[42] More disconcerting, however, is the long and deep history of ecological thinking in Nazi ideology.

German racial purist and virulent nationalist Ernst Moritz Arndt (1769–1860) is cited as the forerunner of modern ecological thought.[43] The hatred Arndt had of the Jews, French, and Slavs melded with his passion and defense of the German environment. Ernst Haeckel (1834–1919), a German naturalist and proponent of natural law, is the one credited as being the founder of "ecology."[44] His views on a hierarchical and centralized social order, euthanizing to achieve racial purity, and animals as moral beings directly influenced Nazi legislation and policy.[45] Daniel Gasman has explained that "Haeckel contributed to that special variety of German thought which served as the seed bed for National Socialism. He became one of Germany's major ideologists for racism, nationalism and imperialism."[46]

Arndt's student Wilhelm Heinrich Riehl correlates a distinctly racialized notion of German identity, agrarian life, and German wilderness. In his essay "Field and Forest" (1853) he states, "In the destruction of the contrast between field and forest you are taking a vital element away from German nationality . . . The German people need the forest as a man needs wine . . . If we do not require any longer the dry wood to warm our outer man, then all the more necessary will it be for the race to have the green wood, standing in all its life and vigor, to warm the inner man."[47]

This sort of romantic natural mysticism, expressed through a deep appreciation for the forest and one that became the frame through which to articulate Germanic identity, repeated in the work of Franz Heske, who, in *German Forestry* (1938), wrote, "German culture sprang from the forest. It is a forest culture. In holy groves the ancient Germans worshipped their gods . . . In the old forests, the present generation seeks to recapture that reverential awe which is the foundation of morality."[48]

Robert Lee and Sabine Wilke have explained how the *völkisch* reverence for "forest feeling" was deployed to full effect in the National Socialist *Volk* film *Ewiger Wald* (Enchanted Forest, 1936). The film was commissioned by Alfred Ernst Rosenberg, a Nazi theorist of the Holocaust.[49] Several authors have drawn attention to the intersection of nature mysticism, racist thinking, romanticism, pantheism, and German National Socialism.[50] George Mosse has meticulously documented how this intellectual history influenced and shaped the German *völkische Bewegung* (people's movement), which later became institutionalized by the German Youth Movement, the Pan-German Association, the Farmers' League, student corporations, and even by German textbooks valorizing a romantic return to an idyllic rural past. Eventually these ideas were incorporated into the German National Socialist Party and Nazi worldview under the formidable leadership of Adolf Hitler (1889–1945).

Völkisch thought advanced a romantic view of Teutonic agrarian life, one uncontaminated by urban life, materialism, and modernity. The Nazis held *völkisch* legend up in stark contrast to scientific rationalism and intellectualism, both of which they attributed to the Jews. Mosse philosophically contextualizes the concept of *Volk*, clarifying that it cannot be literally translated as "a people"; instead, it is a patriotic notion that refers to a unified Germanic people, one that enjoys a transcendent essence. Reclaiming a *völkisch* spirit allowed Germans to reclaim their special connection with nature and become unified as a people. Indeed, the Nazi Rosenberg wrote in his diaries that "the Volk and the Reich must form an inseparable unity."[51]

Stanley Payne has commented, "Notions somewhat similar to *völkisch* attitudes might be found in varying degrees or formulations in nearly all countries undergoing the changes associated with modernization, but only in Germany did *völkisch* culture achieve a broad following both among part of the intelligentsia and the middle classes."[52] Less convinced than Mosse or Payne on the causal connection drawn between *völkisch* ideology and German National Socialism, Petteri Pietikäinen has maintained that *völkisch* ideology influenced only how "*some* Germans perceived reality and made claims about what constituted the true social and moral order."[53]

Hitler might have disagreed with Pietikäinen. He credited German National Socialists with the popularization of *völkisch* ideology. "Not until the work of the German National Socialist Labour Party had given this idea a pregnant meaning," he insisted, did *völkisch* "appear in the mouths of all kinds of people."[54] Mosse has shown that by 1914 the influence of *völkisch* thinking had been so effectively assimilated throughout German society that anti-Semitism was becoming institutionalized. He adds that the "intensity of anti-Semitism can be used to gauge the depth of the penetration of Völkisch ideology."[55] For instance, bluntly connecting race with *völkisch* thinking, Hitler wrote, "The *völkisch* concept of the world recognizes that the primordial racial elements are of the greatest significance for mankind. In principle the State is looked upon only as a means to an end and this end is the conservation of the racial characteristics of mankind."[56]

Robert Pois has gone so far as to describe German National Socialism as a religion of nature.[57] There is a well-documented history of collaboration between German conservationists and Nazis. For example, in 1935 Nazi Party member Hermann Göring (1893–1946) led the passing of Germany's national conservation law, regarded as a literal manifestation of the new worldview under Nazism. Three years later, German conservationist Wilhelm Lienenkämper (1899–1965) published his essay on the "protection of nature from a Nazi perspective."[58] And Reich peasant leader and minister of agriculture from 1933 to 1942 Richard Walther Darré (1895–1953) advocated for the unity of German blood and soil. Blood and soil (*Blut und Boden*) was a racially purified understanding of blood unified under a nationalistic umbrella—the German people—to be nourished by the life-giving force of the soil of which Germans are descendants and a return to the simple and modest values associated with rural life, caring for and living off the land.[59]

Last but not least, Hitler's fascist thinking found expression in his adherence to a strictly vegetarian diet and his deep commitment to animal welfare. The Nazis passed a series of animal and wildlife protection laws that outlawed kosher slaughter, vivisection, all the while introducing new guidelines for humane hunting and euthanasia for suffering animals, along with reports on endangered species and the establishment of nature reserves.[60] Nazi literature includes the following statement: "Animals are not, as before [the Nazi period] objects of personal property or unprotected creatures, with which a man may do as he pleases, but pieces of living nature which demand respect and compassion."[61] Nazi concern for animal welfare combined both the celebration of mankind's animal instincts and a renewed respect for the natural world. Embracing human animal instincts was viewed as another way of bringing mankind closer to nature.

And finally, the public celebration of the Führer as the great protector of animals reinforced the patriarchal figure of Hitler as the father of the German people: "Do you know that your Führer is a vegetarian, and that he does not eat meat because of his general attitude toward life and his love for the world of animals? Do you know that your Führer is an exemplary friend of animals, and even as a chancellor, he is not separated from the animals he has kept for years? . . . The Führer is an ardent opponent of any torture of animals, in particular vivisection, and has declared to terminate those conditions . . . thus fulfilling his role as the savior of animals, from continuous and nameless torments and pain."[62]

ON THE MASS PSYCHOLOGY
OF FASCISM

Why this brief trot through history linking ecological thinking to Nazism and right-wing politics? First, the point is not that environmentalism (and I use the terms *environmentalism* and *ecology* interchangeably here) and fascism are necessarily connected. Like any body of ideas, environmental thought is not intrinsically emancipatory and inclusive, just as fascist ideas are not fundamentally environmentalist. It all depends on how ideas are used and the direction interpretation takes. Second, and not unrelated, the direction any idea takes within a social field depends upon the kind of investment it is given—emancipatory or reactionary.

Even the most liberatory of ideas can become fascistic if the social energies and drives invested in them are reactionary ones. As Wilhelm Reich noted, "The question of '*how*' a new social order is to be implemented wholly coincides with the question as to the character structure of the *broad* masses, the nonpolitical, irrationally influenced working segment of the population. Thus, at the bottom of the failure to achieve a genuine social revolution lies the failure of the masses of people: They reproduce the ideology and forms of life of political reaction in their own structures and thereby in every new generation, despite the fact that they sometimes succeed in shattering this ideology and these forms within the social framework."[63] In order for a truly free social order to come into existence, Reich contended, the authoritarian character structure of the masses had to be undone. In short, the masses need to reach social maturity.

Reich maintained that liberation has to begin by reinventing the structure of the human psyche. It is not enough for the masses to simply desire their freedom, they have to be capable of the responsibility that comes with and leads to freedom. He understood social responsibility as the capacity of the masses to disentangle themselves from "their condition of blind acceptance and craving for authority" in the structures of the patriarchal family, church, disciplinary education, and the all-powerful figure of the nation.[64] All are structured by asymmetrical relations of power, the effect of which, as Surin has noted, is to situate a subject in a "structure of exploitation."[65] And to introduce a theme examined in chapter 8, social maturity will not develop if the narcissistic structure of imaginary identification remains intact.

Rosenberg echoes this point in his assessment of the success of National Socialism in Germany, one that, according to him, had had no precedent in German history. The explanation he provided in 1941 is illuminating: "The consequence of these historic and unprecedented political occurrences is that many Germans, due to their proclivity for the romantic and mystical, indeed the occult, came to understand the success of National Socialism in this fashion."[66] Surin might describe this as an obvious instance of ethico-political subjects being trapped in a "transcendental validation of their subjectivities."[67]

The structure of identifying the cause of socioenvironmental degradation in mystical or apocalyptic terms, then transforming this into an entirely different problem that scapegoats marginalized social groups, that is then used to justify a new social order based upon an essentializing system of natural law is intimately bound up with the machinic problem of interpretation.

According to Gianni Vattimo and Santiago Zabala's account, it is precisely at the level of interpretation that the world will be changed. Presenting another way of understanding Marx's oft-cited claim that the goal of politics is to change the world, not just interpret it, they brilliantly draw out the political dimension of interpretation. Combining hermeneutic and communist thinking, they chart the journey of radical politics, calling for a nonviolent break with the dominant democratic frame. Proposing a discontinuous historical moment, one that inheres in a continuous history but points to another history, Vattimo and Zabala describe this other history as "oppressed history." It is inherently political because it "encourages the defeated and weak to come forward."[68] Alert to the emancipatory potential of an other history, they explain how it "reveals unnoticed possibilities, projects, and rights that were set aside in favor of Western rationality."[69] In this way, oppressed history functions as a minoritarian force. It is a political practice that doesn't depend upon instating a truth in order to work.

Moreover, other histories expose the "'emergencies' of framed democracy" that take place outside dominant truths. They thereby threaten the frame securing the uncontestability of a truth.[70] For example, as I argue in the previous chapter, a state of exception is one democratic frame that presents emergency as a nonnormal situation, legally allowing for the suspension of everyday laws and rules. This is basically what happened in Paris for COP21. Climate protesters were barred from demonstrating in public places throughout the city because France had declared a state of emergency in response to the November

2015 attacks. Also suspended was the deliberative democratic process. As I outline in more detail in chapter 8, the protesters were presented with the problem of how to protest without people. The demonstrators then turned their political weakness into a strength, reinterpreting public space and protest as they performed a protest in the absence of people. Understandably, the weak play an important emancipatory role in Vattimo and Zabala's theory of hermeneutic communism. Characterized as capitalism's discharge and the effect of political indifference, the weak alter the established political order by weakening it.

In the case of the COP21 protest, climate protesters performed their dispossession (of public space and dialogue) in what Vattimo and Zabala might call events of unconcealment.[71] Inserting bodiless shoes, blocks of melting ice, and images of a warming world throughout the city environmental activists performed a collective action that offered forth an alternative interpretation of public space. Instead of a space the state constructs as public through surveillance, privatization, policing, financialization, and management, the activists renewed the public condition of public space as shared and contestable. Disrupting the militarization of public space under state control produced an emancipatory moment that opened "up a new space" for politics to occur.[72]

Hermeneutic fascism puts the other histories Vattimo and Zabala reference to majoritarian use. It relies upon instating truth. Working with an essentialist interpretation of nationhood forged through a subservient connection drawn between blood and soil, an immutable social form is instituted (nationhood is tied to natural law and a racist social order). In this light, the response of national anarchism to socioenvironmental emergency is to submit to a metaphysical position. National anarchism conceives ecological living, and the break with capitalism and state-centric politics, within an already existing order of privilege (natural law), which is then organized by an authoritarian structure (racist nationalism). Although the interruption emergency creates might be welcomed by National-Anarchists, the changes this interruption generates are defensive and violent, because they aim to fend off future disruption and change. For this reason, national anarchism and ecological Nazism constitute instances of hermeneutic fascism.

The growing influence and efficacy of the environmental movement has generated partnerships that cross national borders and with special interest groups that don't advance an especially environmentalist agenda. This is not surprising, because there are a range of environmental concerns and principles that the environmental movement shares with other sociopolitical groups. Indeed, as Douglas Torgerson has pointed out, "contemporary environmentalism . . . took a distinctive form in industrially advanced societies during the late 1960s and early 1970s, amid the emergence of a now familiar range of social movements."[73] Unsurprisingly, the environmental movement shared the normative political language of rights and justice central to the civil rights, feminist, indigenous, gay liberation, antiglobalization, fair trade, labor, landless workers, slow food, and Zapatista movements, to name a few. Like other social movements erupting around the world during the 1960s and 1970s, the environmental movement began as a people's movement; it was not yet an institutionalized phenomenon in the way that it is today.

All social movements rely on developing a broad network through which to spread information, garner support, and build membership. Some, such as 350.org, have a strong charismatic figure, like Bill McKibben, leading the movement. For other groups, for instance the Green Party, leadership is distributed among a variety of political agents. Some set out to influence policy, others seek more systemic change. Andrew Dobson has disparagingly called environmentalism a reformist "managerial approach to the environment within the context of present political and economic practices." The more radical approach, he has claimed, is ecologism, which doesn't believe in a single-issue approach to green politics; rather, it seeks social and political changes that require human beings to radically change their relationship to the world.[74] The dichotomy between working from inside the system so as to change it versus an oppositional politics of resistance is one that also characterizes the conceptualization of political practice—the affirmative politics advocated by Michael Hardt and Antonio Negri on one side and the politics of resistance anarchist groups present on the other.

But why choose? Why not opt for what I have elsewhere called bastard solidarities.[75] I am referring to political theories and practices that engage with and incorporate different political agendas, ideological positions, discourses, and strategies, practices and theories that operate on both sides of the reform-resist and institutional-grassroots divides. Mixing positions and tactics can create important buffers against co-optation as the movement works from within a capitalist system in an effort to change and reactionary radicalization as it strives to revolutionize the structures of exploitation and oppression endemic to global capitalism from the outside.

Rather than neatly and defensively marking out different territories on which to practice environmental politics and identifying whose environmental politics is more or less successful in producing real change, the movement needs to handle political concepts as tools, to borrow an oft-cited Deleuzian mantra, and treat political practices as inherently promiscuous.[76] Basically, the power of environmental politics is less a question of meaning than it is a matter of use value—namely, being used with a spirit of commoning, as I consider in the next chapter. Key in all this is the claim that environmentalism not be used as a cover for political reactionaries or to sugarcoat violent extremism.

5

COMMONISM

One day in the midst of winter, while visiting Detroit, in the southeastern corner of Michigan, I witnessed how still a city could become. Admittedly most cities are hushed by the cold and snow, so what was different about Detroit? The answer arose from the city's sounds, rhythms, textures, and pressure points. These neither built to a crescendo nor were they defined by modulations in key, all of which are qualities I find characterize some other favorite cities of mine: Chicago, Mumbai, New York, Tel Aviv, or Vienna. Life in Detroit was briskly quiet. Feral structures populated the inner-city landscape, contrasting with the manicured lawns of the suburbs. A dissonant song line wove throughout the downtown neighborhoods with their unfilled streets, rusting remains, overgrown lots, and crumbling buildings.

The core of Detroit was virtually gutted at the end of the twentieth century, and by the beginning of the twenty-first century it was quite simply barren. And I don't mean "barren" in the sense that it was an alienating landscape packed with glass and steel, filled with the usual array of architectural monuments to capitalism and economic progress. There was a disquieting bleakness overlapping with the desolation of vacant lots as reminders of this once "Automotive Manufacturing Capital of the World" filled with bustling crowds and a thriving economy remained scattered throughout.[1] Despite the emptiness, this was also "a place of unpredictable encounters," for there were all kinds of sociality occurring among the ecological growth and remaining residents.[2]

The rural-metropolitan-wilderness hybrid central to urban shrinkage directly challenges a common belief that a city consists of a dense concentration of people living in a limited geographical area, one where the primary means of production is nonagricultural. In addition the urban condition of shrinkage tests the dominant current of growth management that has guided urban design, development, and land use. In this chapter I explore how this hybrid presents an alternative to the production and realization of surplus value that predominates throughout the contemporary landscape of neoliberal planetary urbanization. I argue that this process of urbanization is premised upon modalities of urban commoning, or practices that bring a variety of social and environmental struggles into relationship with one another, dismantling the apparatuses of capture that bring land use and the collective energies animating available land, such as communities and ecosystems, under the control of capital.

URBAN SHRINKAGE

Neoliberal policies that led to massive economic restructuring initiatives beginning in the 1970s have resulted in significant population declines for many cities in the West and former East Bloc. As the global market provided new sources of cheap,

nonunionized labor, mass layoffs followed after manufacturing
moved to the global South. When the success of a city no longer depended upon producing goods and products, and if it
failed to quickly reinvent its economic base and identity, it
often entered the hit list of shrinkage.[3] Examples include
Liverpool, which lost 48.5 percent of its population between
1930 and 2002; Halle (Germany), which after reunification
lost 25.4 percent of its population between 1989 and 2003; and
Detroit, which between 1950 and 2004 lost 51 percent of its
population.[4]

In the United States it was the Great Lakes region (commonly referred to as the rust belt and where Detroit is situated)
that was hardest hit by the social and economic changes
brought on by postindustrialization and economic restructuring. Overseas investment in research and development for U.S.
companies approximately doubled between 1985 and 1991, while
in the United States it came to a grinding halt. With the outsourcing of labor, research, and development, the closure of
North American factories followed. For many manufacturing
cities in the United States this resulted in a dramatic decline in
population, high unemployment, and home foreclosures. It was
not only the loss of manufacturing that changed the landscape
of U.S. industrial cities. Suburbanization also led to decentral-
ization trends in the job market, further compounding the
uneven geographies of class, race, and gender endemic to the
U.S. urban landscape. The racial segregation of U.S. suburbia
meant that minority African American populations were further
disadvantaged when it came to attaining whatever employment prospects remained postindustrialization.[5] Together these
factors—postindustrialization, suburbanization, racially skewed
unemployment, population decline, and the subsequent decline
in the tax base—steadily led to the chronic impoverishment of

downtown U.S. rust-belt cities such as Cleveland, Detroit, Flint, Toledo, and Youngstown.

In 1990 unemployment in inner-city Detroit was at 20 percent.[6] The loss in manufacturing jobs disproportionately impacted African American workers, with the firms nearest to the African American population closing or relocating. With the suburbanization of employment during the 1980s in Detroit, it was the "typical black worker" who "experienced a much greater degree of job loss than did the typical white worker."[7] Between 1980 and 1990 black workers in Detroit lost on average approximately "100,000 jobs within a 10-mile radius," while the "typical white worker experienced little job loss."[8] The spatial redistribution of employment to suburban Detroit resulted in "large migrations of the white population from the central city to the suburbs," but the same trend did not occur among the city's African American population.[9] Along with unemployment, the decline in population, and abandoned real estate came increased drug use. Between 1986 and 1988 arrests for narcotics violations throughout metropolitan Detroit grew from nine hundred to sixteen thousand, overrunning city and county boundaries.[10] By July 2013 Detroit had become the largest U.S. city to file for bankruptcy.

A similar story of population shrinkage happening in tandem with growing poverty and unemployment occurred in Liverpool, England. At the beginning of the twentieth century the city had a thriving harbor-based economy built around the railways, the docks along the Mersey, and a steady supply of cheap, unskilled labor. The city was a key ingredient for trade throughout the Lancashire region and the empire as a whole, earning it the title of being second to London in importance—the Second City of the British Empire. In the 1930s it had a population of eight hundred thousand, with the poorest residents concentrated in tiny terrace houses in the districts to the north and east of the city. In an effort to address slumlike conditions, the welfare state constructed public housing on the outskirts, nearly doubling the city's building stock. After World War II slum clearance began, de-densifying and breaking up Liverpool's urban fabric. As in the United States, a series of suburban developments were constructed, replacing substandard housing and basically turning the city inside out.[11] By 1950 the textile industry had slowed and Liverpool's docks had grown quiet, with layoffs following. Between 1978 and 1981 approximately 18 percent of the city's jobs vanished.

Following the onslaught of deindustrialization were the neoliberal policies of Margaret Thatcher, who set out to limit the role of local municipalities and replace them with the private sector. In 1988 Thatcher replaced the Merseyside County Council of Liverpool, and large tracts of public land were sold off to private investors, once more transforming the urban fabric as six to eight units per hectare stood in place of what was once fifty to sixty units. Unemployment continued soaring, and in the 1990s the unemployment rate in the working-class district of Everton was at 44 percent, with one in two households relying upon some form of welfare to survive.[12] The Toxteth riots of 1981, rising crime, and unemployment prompted a series of defensive residential developments to be constructed—cul-de-sac, private housing estates, and extensive surveillance systems—further suburbanizing and de-densifying the city. Today Liverpool's population is approximately half what it was in the 1930s, but unlike Detroit, it has reinvented its identity and turned into an important cultural hub.[13]

It is worth noting that Detroit and Liverpool are just two examples of urban shrinkage that occurred postindustrialization and with the advent of the global free market economy.

There are many others (Manchester; Ivanova, Russia; Ruhr Valley; Hakodate, Japan), which, however, cannot be explored in detail here. Nonetheless, the wave of shrinkage that occurred toward the latter part of the twentieth century is not the end of the story. More recently the real estate crisis of 2007–2012 has seen the shrinkage of postindustrial cities spread to more prosperous corners of the property market, such as suburban development tracts and wealthier corners of the U.S. and European real estate markets.

The U.S. foreclosure crisis began in 2007 because of poor underwriting standards in the subprime mortgage industry (predatory lending practices to nontraditional borrowers) and falling house prices that left homeowners with negative equity. As investors lost confidence in the value of subprime mortgages and borrowers started defaulting on their loans, banks were faced with a liquidity problem as the land and homes they repossessed were worth less than the outstanding loans. In simple terms, financial institutions entered a crisis (symbolized by the collapse of Lehman Brothers in September 2008), causing credit to dry up and the economy to dramatically slow, which in turn led to a drop in consumer confidence and spending, job losses, and the devaluation of assets.

The U.S. Emergency Economic Stabilization Act of 2008 gave the secretary of the treasury the authority to buy up distressed assets in an effort to save the financial and insurance sectors. The US$700 billion bailout was premised on the idea that the corporate sector is too big to fail. Put differently, it was a form of corporate welfare paid for by public funds. Nouriel Roubini and Stephen Mihm have proposed that the bailouts should have been conditional upon a bonus escrow system; this would have deincentivized the risky behavior arising from large bonuses made in response to short-term gains.[14] In a similar vein Nobel laureate and former chief economist of the World Bank Joseph E. Stiglitz has advocated the need for more government regulation of markets in order to minimize the risks posed by negative externalities.[15]

Stiglitz has pushed back against the myth that the global financial crisis was an unforeseeable and unpredictable event. The Asian financial crisis of 1997–1998 was, he had said, an unheeded warning sign of what lay ahead—namely, that financial instability would move from the outer rings of the global economy to the center (United States). He debunks the common belief that markets are efficient, explaining that markets do not self-correct, and the existence of negative externalities makes them deeply flawed and inefficient. Roubini and Mihm have also sharply criticized the view that markets are inherently rational and self-regulating. Like Stiglitz, Roubini predicted early on that the world would be encountering a financial crisis that would begin by hitting Wall Street and the U.S. financial sector before becoming global in scope.[16]

By 2008 the U.S. financial crisis had gone global. In Spain foreclosures started among lower-income homeowners then crept into wealthier parts of Spanish society. By 2012 the Spanish upper middle class made up 60 percent of foreclosures in Madrid.[17] Sadly, many Spanish foreclosures included not just young first home buyers but also their parents, who served as guarantors on their mortgages. In the United Kingdom the Council of Mortgage Lenders reported that there were approximately forty-six thousand properties repossessed in 2009. This figure represented an increase of six thousand repossessions as compared with the previous year.[18] Figures released from the RICS *European Housing Review* show that between 2007 and 2012 prices in the U.K. housing market fell overall by one-third. Meanwhile in 2011, Ireland's housing market was one of the

most distressed, with house prices falling approximately 11 percent over a twelve-month period.[19]

In 2008 foreclosure filings in the United States had grown 81
percent, marking an increase of 225 percent on what they had
been in 2006.[20] As borrowers defaulted on their mortgage pay-
ments and lenders took ownership of their properties, urban
shrinkage moved from the rust belt to the cities of the sun belt.
States such as Arizona, California, Florida, and Nevada were
among those hardest hit by the recession. A trail of empty and
sometimes half-built structures perforated the homogeneous
and orderly landscapes of suburban developments.

On August 12, 2010, RealtyTrac published the *U.S. Foreclosure Market Report* for the month of July 2010. There were 325,229
foreclosure filings (auction and bank repossession, default)
reported during that time. This marks a 4 percent increase from
the previous month and a 10 percent decrease when compared
with July 2009. With one in every eighty-two Nevada housing
units receiving a foreclosure filing, that state reported the highest foreclosure rate in the country. It was followed by Arizona,
Florida, and California. Coming in fifth place was Idaho, with
one in every two hundred forty housing units receiving a foreclosure filing. The report noted that more than 50 percent of the
nation's total number of properties that had received foreclosure
notices came from five states. Following in the order from
highest to lowest, these include California, Florida, Illinois,
Michigan, and Arizona. Despite California's filing notice dropping 38 percent from July 2009, the state still reported that
66,910 properties had received a foreclosure notice in July 2010.[21]

Dairy cows lingering alongside the largely vacant Charter
Pointe suburban development in Idaho signal a dramatic change
in the suburban form of the United States. It appears the era of
the American Dream imploded as single-family homes with

white picket fences, neatly cropped front lawns, two-car garages, and backyard swimming pools were vacated, boarded up, raided, and filled with weeds and overgrown lawns. Basically, the collapse of the real estate market left many U.S. suburbs in ruin, and at the time of this writing the future of McMansion subdivisions and low-density suburbia was looking grim indeed. The global financial crisis resulted in entire urban neighborhoods being littered with deteriorating, abandoned, or half-built homes. This set in play a vicious cycle as municipal revenues dropped along with the falling property values. As neighborhoods became distressed by urban shrinking, the question of how to revitalize the communities left behind grew increasingly urgent.

And so we return to the opening scene of me standing alone in the middle of an empty downtown residential street in inner-city Detroit with boarded-up, burned-out houses in front and behind me and the realization that what I was encountering was life in slow motion. Yet regardless, or should I say in spite of this, there was an astonishing ensemble of forces: vibrant pockets of wildlife married with abandoned industrial areas and reclaimed deserted structures that gracefully stood in defiance against the encroaching emptiness. Ecology, that untamed dimension of material life in the raw, was returning here, slowly claiming the asphalt, inhabiting the buildings left behind, and repossessing private property on its own terms. The remaining residents were pushing back against the drug dealers, meth houses, crime, unemployment, and hunger that had decimated their vacated neighborhoods, adopting the overgrown lots and putting them to work growing food, developing youth job-training programs, providing support for recovering drug addicts, and generally developing community well-being. Put differently, this was another kind of sociality that was also working with the local skills, resources, and ecological systems: urban commoning.

The concept of urban commoning, as I develop it, is not the same as the concept of social capital. Indeed, it is a retort against the theory of value that social capital is premised upon and advances. Contrary to social capital theory, the concept of urban commoning intends to describe modes of sociality that offer alternatives to the production and realization of surplus value, modalities that I insist overdetermine the current situation of neoliberal planetary urbanization. I propose that in order to combat the widespread commodification and privatization of the urban commons we need a clear understanding of how the urban common(s) acquires value, and then from there we are in a better position to realistically examine where and how other options for urbanization might be brought into operation.

NEOLIBERAL PLANETARY URBANIZATION

Accompanying changes to the social and economic landscapes of shrinking cities are material changes to the urban form and fabric. This is not only because real estate values fall, the tax base empties out, and both public and private investment drops; it also signifies a new kind of urbanization that does not revolve around the production and realization of surplus value. This is a slow, destabilizing, green, empty, quiet, and part-time urbanization process. It produces a pockmarked urbanscape that punctures the boundary between urban and rural forms. It is an urbanscape of differentiating scales and temporalities, staggering between spatial clusters and open spaces, with its expanding wilderness swaying alongside and throughout the built environment. This process of urbanization, characteristic of many old industrial epicenters of the West, is also a feature of the recent urbanization of foreclosure. Both conditions have interrupted the dominant urban process of producing and realizing surplus.

In *Capital* Marx explains that value is created through the production and exchange of commodities (today these include assets, products, and services). Value, he explains, is quantitative and qualitative; it can be measured such as in an amount of money; and it is socially expressed through the system of exchange, such as through the trading process when commodities are compared and contrasted with one another.

How useful a commodity is or how much it satisfies a need is what Marx referred to as a commodity's use -value. The material form of a commodity—its use value created by human labor—can be immediately ascertained by studying how much of a human need it satisfies. Surplus value inheres in a commodity but is not realized until the commodity is traded, at which point the human labor embodied in the commodity is abstracted. He reiterates, "It is only by being exchanged that the products of labour acquire a socially uniform objectivity as values, which is distinct from their sensuously varied objectivity as articles of utility."[22] The labor and resources expended in the production of a commodity create a surplus value that inheres in the commodity. However, surplus value is immanent to the commodity (assets, products, services) only until it is actualized through the social practice of exchange, when the commodity is exchanged and realized as profit, income, or price. Regardless of whether we are speaking of urban shrinkage or expansion, the production and realization of value predominates.

David Harvey has succinctly explained as follows: "Capitalism is perpetually producing the surplus product that urbanization requires. The reverse relation also holds. Capitalism needs urbanization to absorb the surplus products it perpetually produces."[23] Together value production and realization constitute the asymmetrical processes of urbanization, extracting surpluses "from somewhere and from somebody, while control over the use of the surplus typically lies in the hands of the few."[24] Finance, investment, and speculative markets drive the urbanization process by presuming a growing rate of profit; all are premised upon the notion that urbanization will in turn facilitate the rate of profit to grow.

The entanglement of global capital and urbanization constitutes the current phenomenon of neoliberal planetary urbanization. In 1970 Henri Lefebvre predicted our current situation of planetary urbanization, announcing at the outset of *The Urban Revolution*, "Society has been completely urbanized . . . An *urban society* is a society that results from a process of complete urbanization. This urbanization is virtual today, but will become real in the future."[25] Later he announces to the reader that the "concept of the city no longer corresponds to a social object," clarifying that "urban reality today looks more like chaos and disorder . . . than an object."[26] More recently, others have begun contesting the usefulness of the concept of the city. Architect Andrea Kahn has claimed that the "city" as an isolated unit of assessment or a bounded entity is today defunct, for the "urban site is scaled through a set of dynamic functions created by fluid interactions between many differentially extensive processes."[27] Similarly, Andy Merrifield has poignantly described a move "away from the notion of a *city* toward *the urban*."[28] And Neil Brenner has asserted that "the urban can no longer be understood as a particular kind of place—that is, as a discrete, distinctive, and relatively bounded type of settlement."[29]

Ultimately, planetary urbanization is a social relation that organizes everyday life.[30] Adding *neoliberal* to the concept of *planetary urbanization* is merely an attempt to historically situate the contemporary process of planetary urbanization and fend off the urge to naturalize this situation. For Harvey, neoliberalism has given rise to a new "kind of state apparatus," which he aptly calls the neoliberal state.[31] The freedoms embodied in the neoliberal state "reflect the interests of private property owners, businesses, multinational corporations, and financial capital."[32] In other words, the state apparatus serves private interests ahead of the public good. Increasingly urban change is managed through public-private partnerships whereby government entities and the corporate sector work together to distribute the costs and benefits of the urban commons. This is carried out largely through policy, zoning, property markets, speculative investment and finance, and development projects. One clear example of this would be the insidiously institutionalized system of theft, otherwise referred to as accumulation by dispossession, taking place in a "state-finance nexus."[33] The state-finance nexus refers to the complicity between state actors and capitalist class power, whereby the institutions of government advance capital accumulation and the legal system protects and promotes this situation. Thus, the concept of neoliberal planetary urbanization concomitantly refers to the Herculean nature of global capital accumulation and the process of urbanization integral to it, all the while keeping the transformative politics immanent to the concept open, to pose the possibility of an urban society yet to come.

As exchange value is established as a social norm, urban life is captured by capital. Merrifield has his finger on the pulse when he notes, "Never before—even more than in Lefebvre's day—has the urban process been so bound up with finance capital and with the caprices of the world's financial markets. The global urbanization boom, with its seemingly insatiable flows into the secondary circuit of capital, has depended on the creation of new mechanisms to wheel and deal fictitious capital and credit money, on new deregulated devices for legalized looting and finagling, for asset stripping and absorbing surplus capital into the built environment."[34] Urban shrinkage is symptomatic of the "homogeneity and fragmentation" of uneven global development.[35]

For instance, it is worthwhile pointing out that urban shrinkage and expansion are two sides of the same coin. Deindustrialization and the trend to outsource labor resulted in the urban landscape of the global South undergoing dramatic changes. Attracted to the economic opportunities that urban areas offer, cities in the global South have experienced rapid population growth. The "world's urban labor force," Mike Davis has noted, "has more than doubled since 1980."[36] It is no coincidence that statistics Davis cites neatly overlap with a time in human history when neoliberal doctrine has been in full swing. Unable to keep up with the demand for residential land and infrastructure development, it comes as no surprise that urban expansion in the global South has resulted in the growth of vast tracts of shantytowns characterized by poor infrastructure (electricity, water, trash collection, toilets, education, and medical services) and substandard, crowded housing.[37] And while there are many differences between the physical conditions of informal housing and urban shrinkage, there are several characteristics they share. The populations of both experience extreme poverty, impediments to accessing the urban advantage, and social stigma. Even the vacated landscapes of foreclosure have their fair share of substandard housing and infrastructure challenges.

PREDATORY PURCHASING

Regardless of whether we are speaking of urban slums or the feral and foreclosed landscape of urban shrinkage, value reenters the equation through the back door of speculative property investment. As the rate of profit for properties declines, because the surplus value of the property has not been realized or has depreciated when properties become distressed, and land is viewed as unproductive, speculative investment in the distressed properties and available land is based upon the assumption that future appreciation in property values or rental income will be realized.

When surplus exists but is not realized, private investment drops. That is, unless a different way of realizing value reenters the equation. One way this is done is when social relations become the basis for producing and realizing value. Social capital, as Sheila Foster has proposed, creates an urban commons through which the bonds of neighborliness, community activities and revitalization projects, and networks of trust return and turn around the blight associated with "unproductive" land.[38] Collaborative structures managing shared resources (community gardens, community urban farms, or neighborhood-watch programs) are strategies communities use to reinvest in their distressed neighborhoods. This can positively impact the social and economic fabric of a community. Community gardens are one example of a collaborative management structure that relies upon social capital for its success. The positive influence social capital has on the "most disadvantaged neighborhoods" includes "increasing rates of home ownership and reductions in poverty."[39]

For Foster, social capital is a common resource that sustains community well-being and protects environmental quality.[40] It can help in lowering crime rates and positively impact the quality of neighborhood life, helping to turn around poor real estate values. Hence, it can also lower the risks associated with investing in overgrown and available properties. Unsurprisingly, then, the idea of social capital as it relates to improved property values has reappeared as part of the solution to the foreclosure crisis, releasing the forces of speculative property investment that led to the subprime mortgage crisis in the first instance.

The real estate crisis of 2007–2012 in the United States has resulted in a resurgence of speculative investment, this time not for the development of suburban housing tracts, shopping centers, resorts, industrial complexes, or malls. Rather, investors have seen the vacated landscape of urban shrinkage as a good business opportunity and have begun buying up foreclosed properties: "To most people, a foreclosure is a bad thing. To a real estate investor, it can be a goldmine. Sure, nobody wants to see anyone lose their home. And unfortunately you or I aren't in the position to forgive the loans of defaulting home-owners. However, with some practice and lots of know-how, you can learn how to help the homeowner make the best of a bad situation—while earning yourself a nice reward in the process."[41] The real estate crisis has also increased demand for rental properties, providing additional incentives for predatory purchasers. In 2010 the apartment research firm Axiometrics reported that 2010 was the best year on record for landlords and rent prices in fifteen years, with rents rising 4.2 percent for 2010.[42] Europeans fared better than their renting counterparts in the United States given that rent controls in some countries can provide renters with a solid safety net (France, Austria, Sweden, to name a few). Alas, with the financial crisis countries such as Portugal began proposing severe social reforms, one of which is the elimination of rent controls.

In the United States even the public housing market is being privatized. Between 2003 and 2005 in Antioch, California, Section 8 households increased approximately 50 percent.[43] One of the main reasons for this sort of statistic is that under the Section 8 Federal Housing Voucher Program (1983) low-income people are being forced out of federal housing projects in downtown areas and provided with incentives (rental vouchers) to enter middle-income housing on the private market. With the foreclosure crisis in full swing, this has resulted in many federally assisted tenants relocating to the new lower rents on offer in the distressed and foreclosed landscapes of suburbia.

The Brookings Institution *State of Metropolitan America* report of 2010 noted that with "bright flight," urban centers are becoming whiter as the educated younger population wants to enjoy a shorter commute and the lifestyle of downtown living.[44] Meanwhile, the larger metropolitan area of the suburbs is becoming more diverse, with the majority of African American and Hispanic populations now living in suburbia. Hence, on the one hand the suburbs might be emptying out as a result of the economic meltdown, but what was once a racially and economically homogeneous landscape is turning into an affordable housing option for others. The single-family housing typology is changing as cheap suburban rentals offer options for multifamily housing. The hope is that with this trend, suburbia could very well become denser and the demographic could permanently shift away from the white nuclear family model to include different kinds of household types, ethnicities, and racial groups.[45]

Yet the social differentiation of suburbia is not necessarily economically redistributive. In the absence of new infrastructure, especially social services and a strong public transportation network, the new suburban poor might diversify the demography of the suburban landscape, but they remain economically vulnerable. The risk is that they will be trapped in a poverty cycle on the urban fringes where few employment opportunities exist. If residents can't afford increasing gasoline prices, or, worse still, an automobile, their geographical paralysis might quite simply render them invisible. This kind of socioeconomic isolation could be far worse than living in the downtown destitute areas they left behind. The point being, viewing the diversification of suburbia as the solution to the foreclosure crisis merely displaces the deeper problem of surplus realization as it relates to inequity and urbanization and the power relations in which they are ensnared. The same can be said for the privatization of vacated downtown areas. Privatization of tax-delinquent land might remobilize the devalued surplus of vacancy and produce a new space for capital accumulation and circulation, but it fails to critically address the speculative dimension of surplus value. One important lesson from Marx's theory of value is that value cannot simply be translated as profit, because the surplus value contained in a commodity is not always realized as profit; sometimes it is devalued. Therefore, surplus value is necessarily speculative.

Foster, in her study of communities transforming abandoned land the city owns into a common resource—social capital—that was being threatened to be sold off to private developers, highlights the ways in which private markets and speculative forces carry more influence on land-use decisions in the United States than do community organizations and "public deliberative processes that consider the larger social, economic, or environmental" issues.[46] She is clearly contesting the neoliberal logic of trying to attract private investment as the solution to urban shrinkage. Her study mounts a formidable case in favor of revising law and policy so as to better account for a community's social capital by recognizing the "inherently public nature" of land used by a community to promote public well-being.[47] She argues that law and policy include within their scope a revised definition of property rights, one that recognizes a "limited type of property interest" that rests upon the creation of common resources—improved social capital, lower crime rates, and environmental well-being. Her solution is helpful and is certainly a step in the right direction.[48] Unfortunately, the social capital generated by the community gardens also attracts the interest of predatory investors.

Community urban agriculture in downtown Detroit caught the attention of millionaire financier John Hantz, who, in consultation with the Kellogg Foundation, saw a tremendous business opportunity in turning vacant city-owned downtown lots into "the world's largest farm"—Hantz Woodlands. The aim is clear: create land scarcity where there currently is none and drive up its value along with producing food for the market. Hantz intends to invest thirty million dollars of his own cash to beautify the downtown fields of Detroit with orchards and develop for-profit farms, maximizing the productivity of the small plots with the latest in green technologies (compost-heated greenhouses and hydroponic growing systems). In order to realize his vision, Hantz requested the Detroit Economic Growth Corporation give him tax-delinquent land in addition to the land he purchases (at below-market rate). In this instance, the value of the land does not equate with its price. This roused the concerns of Detroit's community agricultural sector, which understandably viewed the proposition to commercialize Detroit's urban agriculture as a mask for a massive land grab.[49]

Foster's study of the legal frameworks used to impose the law of private property and profit upon abandoned private property that has been used by a community for public benefit speaks directly to the pitfalls of the "state-finance nexus" that Harvey alerts us to. The irony is that by selling off the land to Hantz Woodlands, the city implicitly acknowledges available land can be successfully put to alternative uses, ones that do not rely upon large-scale privatized development. In defense of its decision to sell off a large portion of Detroit's east side to Hantz Woodlands, the city claimed it could no longer afford to maintain the vacant lots or the loss in tax revenues. But there are other ways the city could have responded.

Garrett Hardin's influential essay "The Tragedy of the Commons" describes a scenario where shared resources are depleted because of overuse.[50] Basically his thesis is that individuals do not act in the interests of the collective but out of self-interest and in isolation to one another. The commons dilemma arises because the use or abuse of the commons by one individual or group negatively impacts the ability of another individual or group that has a shared claim in the resource to use it. The dilemma creates a political problem because if left unregulated, individual stakeholders will overuse and deplete the commons; this, along with their competing claims over a shared resource, can lead to conflict.

Since the publication of Hardin's essay theories and practices devoted to the protection, maintenance, and management of the world's commons have begun to gain traction. In particular studies on the commons were pushed into public prominence in 2009 when political scientist Elinor Ostrom was awarded the Nobel Prize in Economics for her groundbreaking work on the management of common-pool resources (a natural or human-made resource where it is difficult to exclude another person from using the resource and where one person's use of the resource diminishes another's use).[51] Her research rebuffed the popular belief that people cannot work together to sustainably and peacefully manage resources they hold in common and that management of the commons is best left to government or big business. Ostrom's experiments on the different ways people act based on variable incentives led to the important conclusion that individuals are capable of collaborating and developing reasonable rules and systems of self governance that can efficiently, sustainably, and equitably manage resources they share and depend upon. The key to success is the creation of institutions that facilitate and incentivize cooperation among different stakeholders so that they collectively manage common-pool resources for their mutual benefit. The problem of free riders that Hardin has drawn attention to can be overcome by simply improving the lines of communication between people and getting community members, not an official external to the community, to monitor one another's use of the resource in question. Ostrom has identified trust and successful communication between stakeholders as two primary conditions of a common property institution that leads to successful and equitable outcomes, and it is these principles that are at the core of Cuba's experiments in urban farming and alternative land-use policies.[52]

Urban farming and land-use policies in Cuba have supported grassroots labor and the use value of land for the production and equitable distribution of food. In this regard, it offers a helpful framework for developing alternatives to neoliberal practices. Cuba combines the autonomous activities of communities with a heteronomous model of institutional support. The country has successfully supported community urban agricultural activities by connecting the coherency arising from a centralized system of management and strong policies with the decentralized activities of urban agricultural representatives from the Ministry of Agriculture working with communities and individuals.

After the fall of the Soviet Union (1989) and the subsequent termination of the Soviet-led Council for Mutual Economic Assistance, of which Cuba had been a member since 1972, the country was confronted with a food crisis. All of a sudden petrochemical-based fertilizers and pesticides, along with agricultural machinery, were in short supply. Not to forget the partial blockade against Cuba at the time and the imminent threat of a complete blockade adding to Cuban anxieties over food supplies. The country responded by instituting a system of organic urban agriculture called *organopónico* that, as economist Sinan Koont's extensive research on Cuban urban agriculture has shown, has since gone on to "become one of the mainstays of vegetable cultivation in Cuban urban agriculture."[53] The *organopónicos* creatively blend food autonomy, land use, and economic policies with small-scale community-driven agricultural activities, alleviating Cuba's dependence on petrochemical-based agricultural inputs in addition to providing healthy, affordable food. This is because small-scale farming initiatives do not require large machinery, and producing food close to densely populated areas minimizes transportation. Koont reports that more than 350,000 new jobs have been created, and with the total workforce of Cuba reportedly at 4.8 million in 2005, urban agriculture "makes a significant contribution to the country's total employment."[54] Over time, yields have significantly increased, from 1.5 kilograms per square meter in 1994 to 25.8 kilograms in 2001.[55]

Cuba's model of community urban agriculture as integral to government policy has influenced Venezuela's Agro Ciudad movement. Since 2009 Agro Ciudad has transformed more than twenty-six million square feet of available urban land to agriculture.[56] Venezuela developed its urban agricultural program as part of that country's road map to food sovereignty. Food sovereignty, as defined by Vía Campesina, is "the ability of each state to define its own agricultural and food policies, according to the goals of sustainable development and food security."[57] The oil crises of 2002 and 2003 and the global food crisis of 2009 directly posed a threat to Venezuela's food supply. The government responded by diversifying its agricultural economy with a view to enhancing "human well-being, the environment, and culture" through family or collective agricultural activities. It emphasized "endogenous development"—development from within communities and based on local skills and resources. Food is produced in schools, universities, by cooperatives, neighborhoods, and families. It is sold at local markets, and as intermediaries are cut out of the equation, the price of food is lowered. Between 2009 and May of 2012, urban food production in Venezuela grew from 457 tons to 6,340 tons.[58]

The collaborations described in the preceding paragraphs generate an alternative value to the law of value that subsumes all social relations under the umbrella of capital accumulation and surplus value. They are therefore successful precedents in urban commoning. This brings the discussion full circle. How might an alternative value to that of surplus value be generated? Can this provide the basis for a different kind of urbanization process, one that is not already co-opted by the form of value? I think the answer to these questions could lie with processes of urban commoning. The development of the value form of the urban commons that community activities like farming yield raises the thorny issue of how the urban commons has come to be exploited and appropriated, and, in turn, how this reconstitutes the common as noncommon. This problem requires commons scholarship shift from a study of decentralized systems of management with specific policy implications to a deeper philosophical problem concerning the manner in which the commons is being reconstructed as an untapped market opportunity with a view to resuscitating the relative autonomy of commoning as a practice that ruptures the machinations of capital accumulation and circulation from inside, all the while presenting an unmediated outside to it. This is exactly what the Venezuelan and Cuban experiments in community agriculture achieve.

The strength of the Cuban and Venezuelan examples of organic urban agriculture comes from how government policies were harnessed to support collaborations between people, land, and ecological systems—training, marketing, local markets— such that the autonomous activities of the community pervade the institutional arrangements they work in. Instead of simply being an apparatus that captures these energies and places them in the service of capital accumulation, as the governing authority of Detroit has done, the Cubans have developed a more collaborative model, one that challenges the large-scale corporate system of monoculture and the privatization of land and land use. Concomitantly the Cuban situation presents us with a glimmer of hope as it successfully produces food for people, not the global market (cash crops, speculation, fluctuating prices, and for-profit biofuel crops). As the Venezuelan government has stated, it "has a social vision of the food sector: food is not merchandise, but rather, a fundamental human right."[59] Integral to both the Cuban and Venezuelan projects are practices of urban commoning.

SELF-ORGANIZING AND
FUTURE-ORIENTED PRACTICES

Michael Hardt and Antonio Negri share with Ostrom a suspicion toward Keynesian or social democratic ideals of regulation and reform, as well as toward the regimes of private property and law in operation throughout capitalist society. Unlike Ostrom, their concept of the common is not that of pregiven commons; rather, the common is socially produced through desires, well-being, cooperation, and generosity. "By 'the common' we mean, first of all, the common wealth of the material world—the air, the water, the fruits of the soil, and all nature's bounty . . . We consider the common also and more significantly those results of social production that are necessary for social interaction and further production, such as knowledges, languages, codes, information, affects, and so forth."[60] For them, the common is a social relation internal to capital; it is both a relation that capital produces and one that the political endeavors of the multitude bring into existence in its struggle against contemporary regimes of power.

Hardt and Negri have observed that with post-Fordism the production process moved from a quantitative system of material production to a qualitative system of immaterial production that involved the reproduction of information, communication, knowledge, and affects (emotions, excitement, joy). They go on to argue that the common arises from social immaterial production; it is not given in the way that Hardin's commons dilemma posits, for which the institution of private property promises to solve. By substituting "the common" in place of "the commons," they interrupt the collective versus individualistic paradigm informing a great body of scholarship devoted to the study of the commons that presupposes the collective spaces of the commons were privatized by the institution of private property. This is because the common as they define it is neither public nor private; it neither comes under the control of the state nor is it reduced to the atomized world of individual property rights.[61]

Hardt and Negri push the discussion of the commons out of the dualistic framework of public versus private onto a productivist terrain that recognizes the common as the effect of sharing and cooperation and the inclusive logic that inheres in any given instance of the common[62] They are careful to point out that the immanence of the common is not grounded in "any faith in the immediate or spontaneous capacities of society. The social plane of immanence has to be organized politically."[63]

Urban areas are one of the primary sites of common production. Cities are, as Hardt and Negri, have observed, "to the multitude what the factory was to the industrial working class."[64] They are the product of everybody who lives, works, and visits there. And just as the factory was once a site invested with capitalist relations of production and also a site of resistance for labor, Hardt and Negri argue, today it is the metropolis that produces both the conditions of exploitation and political resistance. If the city is both the factory through which capital and the common are produced, then the antagonistic relationship the common poses for capital appears to be less a question of recognizing and organizing the common, as Hardt and Negri contend, than one of activating it, or what might otherwise be called commoning—reclaiming the commons from the process of capital accumulation. Perhaps the *productivist* focus of their analysis—the city as a factory—needs to be expanded to include a consideration of how urbanization *realizes* surplus value and in turn how this extends beyond the finite boundaries of individual cities to encompass a global process of neoliberal planetary urbanization. The second part of this equation is helpful when we try to assess the political potential of commonism.

Commonism produces positions of relative autonomy to the heteronomous machinations of capital by creating alternatives to surplus value that directly engage with how surplus is realized as profit. It also generates a locus of general social value through which other social actors can collaborate in their struggles against the value form of capital accumulation. Detroit's community urban gardening programs produce an alternative to the law of value as realized through exchange and the generation of profit.

The Detroit Black Community Food Security Network, for example, was involved in researching equitable access to healthy food and employment opportunities, work that later formed the basis for the existing Detroit Food Policy Council. This kind of initiative has transformative aims: to create an autonomous food system, one that challenges the corporate model of industrialized monoagriculture operating on a large scale for the global market. It builds coalitions with governmental and community leaders, changing the power relations endemic to the law of value. The Detroit Black Community Food Security Network links the struggles of population shrinkage and available land to other struggles over unemployment, outdated skill sets, crime, drugs, food deserts, and hunger. The common produced through these instances of commoning are also ideological in that they successfully expose the materiality of poverty and urban blight that the commodity form of labor conceals from view.

Commonism creates situations of radical alterity from within the privatized landscape of neoliberal planetary urbanization. It mediates between the natural common of ecological growth, food production, and the artificial common of innovative ideas and collective practices, distributing the benefits, resources, and opportunities that arise from the urban common in support of collective well-being. Simply put, the community farms feed the hungry, support drug-rehabilitation programs, maintain available land, provide training and job opportunities, and, more importantly, they provide tangible collective responses to the scarcities arising in tandem with urban shrinkage. In contrast, the speculative land practices of Hantz Woodlands are premised upon creating scarcity for future gains.

The responses to urban shrinkage described in the preceding highlight a struggle that is ensuing between processes of *commoning* and neoliberal planetary urbanization. The foreclosed and feral landscapes of urban shrinkage transform the rule of property, privatization, and individualism undergirding neoliberal urbanscapes. However, the forces of capitalism pick up the loose threads of the fraying exterior, weaving these common spaces and times back into the interior fabric of planetary urbanization and global capital by subsuming them under the law of value. In this way the common is continually being rendered noncommon, and it is here that the modality of commonism struggles to reclaim the common. In this context commoning announces a shared situation; it is an autonomous political movement. It disrupts the dominant political form that represents the political in terms of a neat contradiction between rational and irrational, order and chaos, state and people, capitalist and worker, a representation that in turn renders the transformative goals of political activity manageable.

Commonism refers to three processes working in tandem. The first is a political project that seeks to construct coalitions between individual, local, regional, national, and even international struggles so as to provide the groundwork for an expansive and wide-ranging politics to form, a politics that aspires to bring about a change in oppressive relations of power. The second is an urbanization process that constructs alternatives to the production and realization of surplus value. The third is collaborative activities involved in concretely transforming the system of exclusive ownership that renders the common noncommon. When taken together these three aspects of commonism strive for human emancipation and environmental well-being. Commonism contravenes the law of value so that exchange value no longer predominates over the land; rather, use value that pervades the land and thereby creates a "new land."[65]

The struggle of commonism is less concerned with finding common ground or the Marxist imperative of organizing the oppressed against their oppressors than with staging experiments with and presenting alternatives to the law of surplus value driving neoliberal planetary urbanization. Commoning actualizes that shared time of futurity so as to open up the present to the difference that inheres in it. Without this, admittedly utopian, form of the commonism experiment, there is little hope that social life can and will change.

In the next chapter I return to the rural-metropolitan-wilderness hybrid that has framed this discussion of commoning. However, in the next chapter I trace how this hybrid appears, or, more specifically, doesn't appear in development discourse and practices.

6

WELCOME TO THE DARK SIDE
OF DIGNITY AND DEVELOPMENT

I love you," he tenderly hummed, losing his balance along the edge where the dirt path met the dusty road. Our eyes briefly touched. I stopped. My friend, who lives here, in the Dago-retti slum of Nairobi, urged me on. The stranger took a long, deep breath, filling his belly with air; then expelling it in one fell swoop, he spat, "and . . . I HATE you!"

Straightening his shoulders, neck, and spine, he wobbles away. My friend nervously dismisses his words. "The unsteady sermons of drunken shame," he explains.

I remain unconvinced. No amount of alcohol defrays saying it the way it is. I am a white female foreigner in his neck of the woods. I bear the official title of a UNESCO Water Chair. I am a professional working at a U.S. university. Despite visiting this Nairobi slum for four years, documenting the water and sanitation services (or their lack), or the warm welcome of friends who live here, and the Kenyan name I have been given—Wanjiru—I am an outsider. My presence here activates a colonial territory of racial privilege and power from within his territory. The marking automatically institutes racial inequity and it energizes a traumatic unconscious. I know it. And my lover-foe knows it.

This encounter between love and hate is constitutive of a racist structure that underpins the traumatic experience of colonialism, as Frantz Fanon so brilliantly described it in *Black Skin, White Masks* (1952). The revolutionary spirit of radical love meets a hate-induced trauma of colonization. On the one hand there is a love that sets a body free as a subject of political change. On the other hand, there is a hate that imprisons the body as an object of normalization, naturalizing it with exclusionary distinctions.

What interests me is how sustainable development in a neo-colonial context can co-opt emancipatory struggles by turning what is otherwise a political matter of black or brown liberation into a moralizing consciousness of individual dignity and self-worth. "Moral consciousness implies a kind of split" Fanon wrote, "a fracture of consciousness between a dark and light side. Moral standards require the black, the dark, and the black man to be eliminated from this consciousness." For this reason, the black man "is constantly struggling against his own image," Fanon explained.[1] Hounded by the moralizing value of realizing self-worth and slowly destroyed by the financial mechanism of debt, the dignity-development nexus can prioritize dignifying individual slum dwellers ahead of a political project that addresses structural forms of violence.

The material struggle between love and hate is *the* defining problem of international development, and to suggest otherwise is shockingly naive. It is tantamount to claiming the racist legacy of a colonial past has somehow been miraculously expunged with the formation of the independent Republic of Kenya (1964). At worst, such a claim is complicit with colonial violence, the manner in which it skewed the distribution of Nairobi's land resources through a racist spatial hierarchy and the "poverty of power," as Ananya Roy so fittingly calls it.[2] Further quoting

Roy, this struggle also addresses "how the problem of global poverty is entangled with the making, and unmaking, of new models of global citizenship . . . [and] those empowered to imagine the persistent invisibility of proximate poverty."[3]

Let me start with a brief exercise in material mapping. Having started out as a railway supply depot in 1899, Nairobi later became the capital of British East Africa. The population of Nairobi is estimated to have been 8,000 by 1901. By 1948, this figure had grown to 118,000, jumping to 343,500 by 1962. Under the colonial administration Nairobi was divided into European, Asian, and African (migrant workforce) residential districts. Travel to the city by Africans was restricted. After independence this restriction was lifted, and Nairobi's population increased dramatically. What had been a population of 350,000 at independence quickly grew to more than 1 million by 1980, 2 million by 1990, and, by 2014, the city's population surpassed 3 million.[4]

The majority of urban growth has been in the form of unplanned settlements. In 2009 more than half the population was living in informal or slum settlements.[5] A slum household is defined as a group of individuals who live without one or more of the following: (1) access to improved water, (2) access to improved sanitation, (3) sufficient living area, (4) durability of housing, and (5) security of tenure.[6] The Kenyan Ministry of Housing reports that 89 percent of the urban population in Kenya cannot afford a mortgage, and despite 83 percent of the country's housing demand coming from low-income families (the majority of Kenyans), 80 percent of housing is built for the country's high- and middle-income population.[7] This is embodied in slum subjectivity.

In the slums the rainy season quickly transforms dusty alleys and roads into disease-infested streams, fouled by chemical and biological waste. Open drains carry water contaminated with biological matter and industrial waste. Children play in and along the edges of water channels. Flimsy dwellings are constructed using sticks, mud, trash, or corrugated iron. The pigs, chickens, and goats that humans eat roam in and out of shacks, wandering the land in between, eating the debris of human, animal, and industrial waste.

Homes are overcrowded, and cramped living conditions both inside and outside provide fertile ground for epidemics to rapidly spread, such as the massive outbreak of cholera throughout Kibera slum in 2015. The more it rained, the more the Kenyan health authorities scrambled to contain an otherwise curable bacterial disease. This is just one example of a myriad of environmental hazards slum dwellers face on any given day, hazards that climate change is compounding.

Add into this mix the cumulative health challenges arising from years of navigating a polluted environment; the flux and flow of people, goods, services, animals, water, air, and waste; inadequate housing; poor infrastructure; the movement and accumulation of fluids and solids; insufficient income; and food insecurity. This situation is what the former executive director of the National Association for the Advancement of Colored People Benjamin Chavis has called "environmental racism."[8] Although Chavis was referring to black communities in the United States bearing disproportionate environmental burdens because of the placement of toxic and hazardous industries in their communities, the unhealthy living and working environments of Nairobi's slums are historically connected to racially motivated urban policies under colonialism that continue to materialize in the slums to this day.

Not everyone living in the slum faces the same level of risk: women and children are more prone to suffer from poor health than men; inadequate sanitation is directly correlated with higher absenteeism for girls in school; women are more vulnerable to being assaulted than men when using public toilets' and the economic burden of health care and medical costs impacts women's income more than that of men.[9] Recent migrants to the slum are more disadvantaged than their intergenerational neighbors; they live in more precarious parts of the slum (where flooding is greatest), in flimsier housing, often paying higher rents and experiencing greater housing insecurity, along with fewer employment opportunities.[10] In other words, a slum is not a uniform concept. It is a highly complex state of affairs calibrated by geography, history, social norms, cultural attitudes, and economic activities. This situation demands we reconceive how dignity is used in development discourse. For when development agencies set the goal of producing dignity in individual subjects, they flatten urban life in ways that mask the variegated ways in which bodies, land, resources, built structures, animals, finance, and energy mutually constitute one another. For this reason, *how* dignity is employed as a platform for sustainable development demands urgent reevaluation.

When UN Secretary-General Ban Ki Moon makes a "call to dignity" around which the United Nations and the world "must respond" with all their "vision and strength," what does this actually involve?[11] How does the advancement of dignity play out in concrete terms? Is dignity a universal value shared equally by all, or a relative concept that changes according to cultural and circumstantial differences; or can it be both universal and particular without sacrificing its coherence?

The goal of this chapter is to resituate the moral discourse of dignity as a human right within a political discourse of liberation and empowerment. This requires dignity be conceptualized relationally instead of as an individual property right. Relational dignity involves a dialectical movement between the universalism of a shared right and the singular realities of processes of materialization. Relational dignity consists of a variety of spatial configurations, and it activates the revolutionary potential of futurity and past conflicts to inform present-day struggles. Dignity, I argue, is an inclusive social relation that expresses itself in the very forms that make up its content.

HISTORY OF DIGNITY

To begin, the central place dignity occupies in development discourse and policy leans on a long intellectual tradition that harks back all the way to Cicero (106–43 b.c.e).[12] When Cicero stated that "sensual pleasure is wholly unworthy of the dignity of the human race,"[13] he instituted a speciesist hierarchy whereby humanity holds a higher status by virtue of our capacity for reflection. Dignity indicates a position of distinction held exclusively by human beings. In time, dignity became an evaluative term, and the connection to status narrowed to refer to a behavior that is worthy of respect—dignified, dignitary—and commonly associated with someone of higher social standing.[14]

It was of course Thomas Aquinas, and later Immanuel Kant, who viewed dignity as an intrinsic value and an attribute tied to the exercise of reason. Aquinas noted that "dignity signifies something's goodness on account of itself."[15] For Aquinas, dignity was tantamount to something occupying its rightful place in God's divine order.[16] According to the theological view of existence, human beings are made in God's image and are an expression of God's creativity, hence people ought to be treated in a manner that is consistent with the special standing they enjoy within a divine order.

This is not to suggest that reason plays no part in the theological history of dignity. Remember that in the *Summa Theologiae* Aquinas clearly connects the special place human beings hold in between bestiality and godliness to a human's capacity for reason. "A man who sins deviates from the rational order, and so loses his human dignity in so far as a man is naturally free and an end unto himself. To that extent, then, he lapses into the subjection of beasts and their exploitation by others."[17]

In a similar vein Kant insisted that "morality is the condition under which alone a rational being can be an end in itself; because it is possible only by this to be a legislating member in the kingdom of ends. Thus morality and humanity, in so far as it is capable of morality, is that which alone has dignity."[18] The theological view maintains that dignity is a form of recognition: recognizing humanity's special status in a divine order where all people enjoy an equal standing by virtue of their membership in the same species.[19]

Kant, too, understood dignity as a form of recognition: to recognize the status of an individual within a moral order, that individual has been socialized and has learned the art of self-control through the exercise of reason. Dignity is a right human beings share in common by virtue of their capacity to be moral agents in the world. Unlike market price, Kant claimed, dignity is not a means to an end; it exists outside the system of exchange, "elevated above any price, and hence allows of no equivalent."[20] Similar to the theological position, then, Kant argued that dignity has infinite value; it is intrinsic and beyond measurement and calculation. The main difference between the Kantian and theological views of dignity is that, for Kant, reason and a moral capacity, not God, are the source of human dignity.

All these views share the assumption that human dignity is connected to the special status humans occupy in the world.

Environmental degradation dramatically challenges this assumption. The recent "Renewing the Earth" statement issued by the United States Conference of Catholic Bishops attempts to amend the analytic emphasis on intrinsic worth historically confined to human beings to include the environment: "People share the earth with other creatures. But humans, made in the image and likeness of God, are called in a special way to 'cultivate and care for it' (Gen. 2:15). Men and women, therefore, bear a unique responsibility under God: to safeguard the created world and by their creative labor even to enhance it. Safeguarding creation requires us to live responsibly within it, rather than manage creation as though we are outside it. The human family is charged with preserving the beauty, diversity, and integrity of nature, as well as with fostering its productivity."[21] This shift in emphasis gives the environment an intrinsic value, and the special role human beings have is to protect all God's creation equally. This view is not entirely antithetical to Kantian thinking, where human dignity is connected to the human capacity for moral reasoning and action.

George Kateb has extracted the environmental implications of Kantian thinking, arguing that human beings hold a "unique capacity to serve as the steward of nature."[22] The uniqueness he refers to is "the will of human stewardship" toward nature, and it is connected to what he calls the dignity of human stature, "the selfless subordination of human interests where need be, and in impersonal appreciation when the stewardship of nature is made a central task of thinking."[23] Stature is a nonnatural human capacity that refers to the infinite creativity of human potentiality as expressed through indeterminately different human cultural activities, of which the stewardship of nature is but one example among many. Revealing a humanist predisposition inflected by Sartrean existentialism, he explains that

when human beings deny their unpredictable creative potential, "they are guilty of bad faith."[24] Stewardship is a "contribution that only humanity can make . . . its stewardship depends
on commendably unique traits and attributes that help to make
human beings partly not natural."[25] Clearly, damaging the
environment would be an example of human inauthenticity
(bad faith), a denial of human agency. In a nutshell, it is self-
deception. On this crucial point, dignity is essentially secular
and existential; it is not a moral value.

Kateb has shown that human dignity is both a dignity that
all humans equally share as a species—human status—and this
comes from the exclusively human qualities and characteristics—
human stature—that are existential values all humans share and
which are individually realized by a person's moral capacity.
But does this view do justice to the horrors of history? Kateb
would argue that if there were only human wrongdoing in the
world, there would be no human dignity or morality to speak
of. For him, the fact there are admirable human traits and
attributes sets the human species apart from other species. And
this is what makes human stewardship of nature not only possible but also a specifically human ability. Actualizing the unique
and commendable traits and attributes of human stature through
the stewardship of nature, people realize their dignity.

The question that arises for me is, doesn't environmental
degradation break apart the presupposition of human totality?
Doesn't the sociality of a slum, for instance, its prepersonal mixing of human and animal excrement with industrial waste, and
the contaminated flows of water chiseling out the ground
between and under flimsy shacks, or the air polluted with fecal
matter and chemicals mixing with water and bodies, along
with the power relations of intergenerational slum resident
status, gendered disadvantage, and the material repetition of a

racist colonial past in the present, together implicate and accordingly demystify the concept of a dignity unique to humans? Doesn't this messy mixture of bodies, matter, energy, and affects make clear that dignity is an impossible albeit necessary concept that animates the transformative agency of political practices, all the while denying dignity its integration in the human form or artifacts? This is a politicized view of dignity. Dignity thus conceived is not another empty exercise of moralizing value that guides and channels the wild desires and energies of the social or an existential humanism that offers a universal and normative view of the human. Rosi Braidotti has poignantly described how humanism institutes the "self-regulating and intrinsically moral powers of human reason."[26] The problem is that a clear delineation is formed between who is considered human and who isn't by relegating "sexualized, racialized and naturalized 'others' to disposable bodies."[27] The discourse of dignity as an individual property right in international development establishes the logic of disposability Braidotti draws attention to.

WHO DOES SLUM DEVELOPMENT BENEFIT?

It is striking just how much attention is paid to dignity as a human right in development discourse and practice. Dignity has become the dominant conceptual frame used in humanitarian outreach, international aid, and the UN development goals, to name a few. For example, as soon as World Bank Group president Jim Yong Kim declares the bank will create "ways for some of the world's most vulnerable people to attain lives of greater dignity and opportunity,"[28] he aligns dignity with development. Obviously he, like many development agencies, is paying special attention to the UN Universal Declaration of Human Rights (1948), which states that "recognition of the inherent dignity and of the equal and inalienable rights of all members of the human family is the foundation of freedom, justice and peace in the world."[29]

Property underlies the dominant conception of dignity, guiding both legal and development discourse. When employed in the legal field, dignity is a property to be protected. Drawing on the legal view of dignity as an inalienable right, development broaches the question of dignity as a property to be imparted or secured. Yet the advantages and disadvantages of development don't necessarily accumulate.

When the development agenda of enhancing dignity is implemented, does it become a tool of oppression or emancipation? Does it facilitate or inhibit a criticism of racism and environmental degradation in the context of global capitalism? For example, what are the consequences of supplying electricity to the slums in Nairobi, or of providing new housing ownership opportunities for slum residents? Slum upgrading promises to improve the lives of slum dwellers by cleaning up slum environments. Granted, small-scale slum infrastructure modifications have been taking place for a while now, but the mainstreaming of slum upgrading on a larger scale brings with it some potentially serious social and ethical consequences. As these promising changes are normalized, it is crucial to critically evaluate not just *who* and *what* is included but *how* they are included in the new developments under way.

The two ambitious goals of the World Bank are to "end extreme poverty within a generation and boost shared prosperity."[30] With more than 80 percent of the world's poor living in developing countries, the World Bank has publicly stated its primary commitment to be the alleviation of global poverty. And yet the majority of the bank's voting rights are held by developed countries; this is even after the World Bank reformed the inequitable distribution of its voting shares in 2010. High-income countries such as the United States collectively hold 60 percent of the bank's voting power. Together middle-income countries such as China, India, and Brazil have approximately one-third. Meanwhile, low-income countries, which, I would like to add, are the primary constituency impacted by World Bank policies and initiatives, wither along the sidelines with a mere 6 percent of the bank's voting rights. The International Development Association of the World Bank is responsible for providing grants and loans to the poorest seventy-nine countries, with a collective population of 2.5 billion. Low-income countries are caught in the ultimate catch-22 scenario: to have voting shares in the World Bank requires a country make a financial contribution to the bank. Low-income countries cannot afford to pay. Changes to this arm of the bank saw a minimal increase of just 3.32 percent in voting power going to low-income countries and a 4.3 percent increase to middle-income countries.[31] This is just one of the many ways racism is geopolitically institutionalized.

Such is the North-South contradiction—for that matter, it is any contradiction arising from severe inequity and the privilege this creates: love entangled with hate, distrust hidden beneath hope, and somberness born from the upbeat rhythms of missionary leftovers. Far, far away, not so far that it remains away, the far and near of colonial trauma call out across time, saturating this moment, right here, right now. Amid the clamor of oppression; the grumbling of hunger and exhaustion; the thievery of land, resources, and labor, where the spoils of colonialism are reclaimed through debt and historical guilt is relieved by "first" world aid. Today, this toxic inheritance is evident in the slums and the institutionalized upgrading programs of the rapidly urbanizing global South, which disaggregates environmental, human, and nonhuman factors from one another.

As Gustavo Esteva, Salvatore Babones, and Philipp Babcicky have reminded us, development discourse and practices assume a model of scarcity when indeed there is more than enough to go around in the world—that is, if the distortions arising from the system of global capitalism are modified and made more equitable. Furthermore, development assumes an underdeveloped or less-developed community and in so doing presupposes a normative ideal that the global North grafts onto the global South.[32] All in all, disaggregating environmental flourishing from the well-being of humans and other-than-human species fails to comprehensively address the complex ways in which autonomy is diminished.

Billed as fostering "A Life of Dignity for All," UN Millennium Development Goals were pegged to measurable outcomes such as Goal 7.D: "Achieve by 2020, a significant improvement in the lives of at least 100 million slum dwellers."[33] In quantitative terms the goal has been achieved well ahead of time with more than "320 million people living in slums" gaining "access to improved water sources, improved sanitation facilities, or durable or less crowded housing, thereby exceeding the MDG target."[34] Too often, though, dignity is pegged to the number of slum homes that have electricity, the number of water and sanitation facilities, or the quality of housing materials used, ahead of a qualitative assessment that focuses attention on living a fair life, food sovereignty, or a life free from oppression, exploitation, and violence. The quantitative indicators of slum upgrading are not value neutral; they replicate a North-South hierarchy that uses the logic of the free market to achieve its goals, without challenging the naturalized view of the power of development. In the process stories of success and prosperity that unfold each and every day in quieter and less-conspicuous ways than building large concrete apartment complexes and sanitation facilities go unnoticed.

How might the statistics tell not only stories of material austerity but also offer narratives of material flourishing? On this point, Arturo Escobar has advised, "One should be able to analyze counting in terms of its political consequences, the way in which it reflects the crafting of subjectivities, the shaping of culture and the construction of social power—including what these figures say about surplus material and symbolic consumption in those parts of the world that think of themselves as developed."[35] All the slums I have spent time in are filled with situations of thriving materiality. Kibera has an active urban farming culture that feeds entire households. Along with the tomatoes, kale, and other greens grown in sacks come the health benefits of a regular diet filled with fresh produce; household incomes are supplemented as extra food is sold; and urban habitats are transformed as the dusty or muddy (depending on the season) common area in the middle of shack neighborhoods are filled with green edible produce. I heard birds in the spaces where farming was taking place, women joyfully singing with their daughters, and the greener the atmosphere the cleaner the air. The power of Kibera's urban farming comes from an affirmative and relational operation of material power engaging light, air, water, food, human and nonhuman animals, music, and laughter. This is a power that cannot be located in an international development agency or a state actor. Nor could we call the power in question repressive because it doesn't operate by subjugating unpredictable materiality to the concrete arm of slum upgrading development in the production of what Foucault called "normalized subjects."

If dignity is broached as an unconditional value intrinsic to the very being of human existence as outlined in the Declaration of Human Rights, then it follows that no number of statistics can fully represent dignity. For this reason, dignity operates like an empty signifier, turning the very problem of what dignity means (more housing units, more toilets, more water taps, more electrical lines) into a symbolic one at best. Nonetheless, as a value that is measured and calculated in support of the UN Millennium Development Goals being reached, dignity is also broached as an empirical and ontological experience, which implicitly recognizes that dignity occurs in history. The latter presupposition promptly exposes the relative and contextual character of dignity.[36] Fredric Jameson has already described this in his criticism of Anglo-American politics: "The method of such thinking, in its various forms and guises," he says, "consists in separating reality into airtight components, carefully distinguishing the political from the economic, the legal from the political, the sociological from the historical, so that the full implications of any given problem can never come into view."[37] This compartmentalized way of knowing the subject of the slum has material consequences. It reorganizes, redistributes, and regulates material flows that submit everyday life to what Foucault called a "regime of truth," which is an effect of power, legitimating the development agenda and invalidating everyday life in the slums. So, how does this play out on the ground?

The upside is that appeals to dignity in the orbit of development have led, over the years, to more inclusive development projects that empower some members and pockets of the underprivileged communities they work with. The increasing focus on projects that are mindful of mainstreaming gender in development are especially encouraging in light of making development outcomes socially equitable. Slowly there has been a shift in thinking away from the neoliberal trickle-down approach to a trickle-up principle that favors grassroots development ahead of macrodevelopment projects that woefully disregard the autonomy of the individuals they are meant serve. The Kenya Slum Upgrading Programme (KENSUP) for water and sanitation facilities in Kibera slum or the new housing compounds constructed on the edge of the Kibera spring to mind.

I conducted site visits to numerous Nairobi slum water and sanitation facilities over a four-year period. They provide not only an important service to the community but also much-needed employment opportunities for women living in the slums. Meanwhile, the new housing is an attempt to move people out of unhealthy overcrowded living conditions into new, high-rise buildings with running water, a toilet, shower, and electricity. If assessed in isolation, these projects are undeniably commendable. The pitfalls come into view when we introduce a relational analysis of how dignity works in the slums and the manner in which social, economic, cultural, political, and environmental conditions violently align.

For instance, one downside with the new sanitation facilities is that people have to pay five shillings to use the toilet and ten to use the showers. To put this in perspective, one hundred Kenyan shillings is approximately one U.S. dollar, and ten shillings is equal to ten cents. Now five cents to use the toilet doesn't sound like much, but if you contract a stomach bug (which many people in the slum suffer from on a regular basis just because of poor water quality and other forms of contamination that come from open drains and sewage), then five cents several times a day for a family of as many as ten people where the average household is living in extreme poverty (US$1.25 a day), then the cost of using a toilet or taking a shower once a day quickly becomes a luxury, especially when the costs associated with meeting everyday necessities such as water (ten to fifteen shillings per gallon), food (fifty to one hundred shillings per day), and rent (fifteen to twenty-five hundred shillings a month) are factored into the equation.

The government and Umande Trust water and sanitation facilities in Kibera are commonly managed by women (I have yet to meet a man who manages one of these facilities) who work from dawn until dark scrubbing the latrines and showers, day in and day out, earning anywhere from a hundred shillings a day to, if they are very lucky, three hundred. The daily grind these women endure is disguised by the development-as-dignity logic. The women who manage and clean the facilities that I spoke with explained that there were few or no other employment options available. Although they were relieved to have a job, they reported not having enough money for the family, long hours without a break, and a general unhappiness about their working conditions. The question is whether a lack of employment choices justifies earning a few dollars a day for twelve to fifteen hours of work. The following World Bank indicator might help put this into perspective: "Extreme poverty is defined as average daily consumption of $1.25 or less and means living on the edge of subsistence."[38]

One serious drawback of KENSUP is the tenuous financial situation endemic to life in the slums. In a strategy document, it clearly states that the KENSUP facility "comprises a small team of specialists on international and domestic financial institutions and financing models. Their task is to seek out and develop mechanisms to mobilise domestic savings and capital for affordable housing, and to liaise between financing institutions and the normative and technical cooperation activities of UN-HABITAT."[39] Katherine Rankin has underlined the predatory logic of microfinance schemes in low-income contexts, explaining how they transform an otherwise supportive social network into a harmful one that not only facilitates development interventions into the community by indebting it but also qualitatively changes social networks, turning an otherwise supportive mechanism into a shaming device.[40]

While the Kenyan government is clearly leading the slum-upgrading program, the project uses Western housing norms such as individual rooms and a privatized understanding of domestic space. In slum homes tenants live in one open area that they divide as they need to, often by hanging sheets over a rope hung through the space, transforming eating areas into a sitting room, bedroom, kitchen, or storage area as needed. This kind of spatial flexibility indicative of the crowded living areas in the slums is more amenable to meeting a variety of collective needs.

In addition, the upgraded high-rise apartment complexes do not have a collective outdoor area that can be used for sack farming, cutting off the possibility for food sovereignty. The vertical organization of the apartments breaks apart the patch-work organization of the slums, which are more conducive to multiple functions. Furthermore, the KENSUP model is built around a neoliberal economic ideal of property ownership and the development of a housing market, commodifying basic services such as water and sanitation and turning insecurity of tenure into a profit-seeking mortgage mechanism.[41] In this way, sociality is privatized as it is transformed into a series of business partnerships; friends and neighbors are mobilized as neighbor-managers and are tasked with collecting debt from members of their community.

International development works on the premise that bodies necessarily conform to the immutable laws of property, finance, debt, a free market, and individual choice. Even the environmentalist rallying cry of collectivization has become a new principle of housing cooperatives that facilitate the move from "slum" housing to the new housing compounds being built on the edges of Kibera. The co-ops oversee the microfinance and credit schemes administered to their slum neighbors, managing savings collection and the repayment of loans. Together these add up to a neoliberal conception of the good life. Just design them a building that has running water, a bathroom, and a bedroom, and informality will be flawlessly formalized.

The take-home lesson: the slum is a form of content. The form of expression for the form of content of the slum is "a lack of dignity." Dignity is, therefore, an action, not just a concept or the definition of a human right. Dignity, or the pronouncement of a lack of it, acts upon the body of a slum dweller, changing the physical environment of the slum, altering how its economy works, and changing its social organization along with it. The form of content and expression reflexively articulate each other.

There is no slum without slum dwellers. They are the substance of the content of the slum. There are also development officers, government officials, visitors (such as myself), animals, deteriorating infrastructure, contaminated water flows, and polluted air that make up the materiality, or substance, of the form of content that is the slum. As a goal of development, the mantra of "developing dignity" emerges as an expression that traverses into the bodies of the slum dwellers, shaping them in the process. As new high-rise housing is built for those same people, new relationships between environment, finance, and bodies are formed. Not all slum dwellers will have access to the new housing. The new housing compounds of KENSUP are for reliable savers who have regular employment and are willing to enter into debt to purchase a home. Here the form of expression (pronouncement of a lack of dignity) becomes the form of content (microfinance, indebted bodies, high-rise compounds, walls lined with barbed wire, massive sewer lines reshaping the waterways through the slum).

The relationship between expression and content is not simply causal (moving in a one-way direction from content to expression). The pronouncement of a lack of dignity changes the bodies of the slum dweller through the cumulative projects that make up the dignity-development nexus: new water and sanitation facilities, work opportunities for women, new sources of gender inequality, the removal of homes, indebtedness, higher housing costs. The political potential of these projects resides in the process of articulation, the moment of passing from one form to another before it has actually striated into another hierarchical social relation (gender inequity, indebtedness, higher rents, and so on). Rather than think of dignity as a property, an inalienable human right, the political power of the concept of dignity resides in considering it as a social relation, as a combination of actions that enhance the ability of all to partake in collective actions (decision making, ecological flourishing, financial autonomy, food and water sovereignty).

Now, the question of dignity is not so much what it means but how it works. How does dignity come into being? At this crossroads I am drawn to a Freudian analysis. For if dignity is inherent to the human being but is believed to be currently absent for some people (in this case poor blacks), then dignity is repressed. This is what the UN development goals set out to rectify. The absence of dignity, though, is symptomatic of a fundamental sociopolitical resistance that censors not just the emergence of dignity but also how it appears—in this case, it makes its appearance through the indebted black body, the gendered body selling its labor power to an entrepreneur, and the transformation of excrement into clean energy flows (biogas). The violence of global capitalism comes from the continual production of an excess; in this case it is the slums. That excess cannot be repressed, so it returns in different form, such as the high-rise apartment complexes paid for by a newly indebted slum resident whose financial backstop is either intergenerational debt or turning into a slum landlord themselves by forcing all members of the family into one room and subletting the other.

THE HUMAN RIGHT TO ENVIRONMENTAL FLOURISHING

Development champions a universal concept of dignity whereby dignity represents a given set of material conditions that are economically determined. This is in stark contrast to how design in the public interest uses the concept of dignity. For public interest design groups such as Kounkuey Design Initiative (KDI), Solidarités Internationale, or the Human Needs Project (HNP), which all work with Kibera communities, dignity is conditional and situated, creating what Roy so fittingly describes as "new imaginations and platforms of solidarity and alliance."[42] In other words, it is denaturalized. Dignity is a transformative élan, a utopian form, that activates activist practices throughout the slum.

In its utopian form, dignity prompts us to aspire to a better world than what currently exists without falling into the metaphysical trap of the future aspirations eclipsing present-day realities. Fredric Jameson once described this as the difference between the form and content of utopia.[43] The political potential of dignity occurs through a radical fracturing of the present, bringing it into a dialectical relationship with its future potentiality, setting out to create a future that is different from the present without idealistically defining what that the future looks like.

KDI works across projects involving people, materials, economics, and ecological growth, creating urban farms using compost made from composting toilets, toilets constructed using interlocking soil stabilized blocks made of mud and a small amount of cement, using them to transform environments from within the slums working with slum residents.

The urban farms of Solidarités Internationale embrace the materiality of life in the slums. Farmers fill sacks with soil, grow kale using a spray concoction made of ingredients they have grown (a solution consisting of chili, onion, garlic, water). The concoction doesn't kill insects but merely acts as a repellent. Plants are nurtured, and some that I have seen have grown to more than five feet in height. The urban farms in Kibera are platforms for food sovereignty, providing high-nutrient fresh produce for free or at a minimal cost to Kiberans and new sources of income for farmers.

HNP tackles the water and sanitation problems of Kibera in a self-organizing materialist way, bringing biological matter (feces and urine), human bodies, water, solar energy, natural filtration systems (sand and rocks) into relation with one another to create a system that produces its own potable water, cleans dirty water, and provides equitable employment opportunities that transform people's lives. All three work with the generative and affirmative potential of everyday life in the slums, creating new ways of connecting matter, bodies, water flows, energy, air, and economics.

Following Braidotti's view of posthuman feminism, perhaps the initiatives described here constitute a form of posthuman environmentalism. In all, dignity is used to dismantle the urban-rural, nature-culture, human-nonhuman, formal-informal divides. Dignity is thereby considered as a form of machinic urbanism, an urbanism that attends to the productive connection of life and how it comes to life. Machinic urbanism triggers activist practices that attend to processes of materialization.[44] For sociality arises spontaneously as communities identify sites for action; articulate collective obligations; work with soil, water, air, food, energy, bodies, and other-than-human species; reinvest in new collectivities; invoke equitable labor models; and create inclusive spaces infused with skill sets and aesthetic sensibilities held in common.

Edward Said summoned a similar concept of dignity when he connected it to activism. For him, dignity set in motion a political commitment, igniting resistance against oppressive and exploitative systems. He stated, "The vicious media and government campaign against Arab society, culture, history and mentality . . . has cowed far too many of us into believing that Arabs really are an underdeveloped, incompetent and doomed people, and that with all the failures in democracy and development, Arabs are alone in this world for being retarded, behind the times, unmodernized, and deeply reactionary. Here is where dignity and critical historical thinking must be mobilized to see what is what and to disentangle truth from propaganda."[45] Said was not suggesting the state should be defending the dignity of its citizens, or that human dignity resides in an individual's capacity to exercise reason and moral judgment. It is striking to learn that Said's view of dignity was not simply an individual respecting another's agency but the means through which agency comes into being.

Shifting the focus from development to design thinking, design in the public interest adopts a contextualized approach to environmentalism, agency, economics, and cultural specificity, working in the middle of these to discover the activist logic of dignity. In this way, dignity makes a singular and random appearance. It is an event, not an attribute that a subject possesses.

Perhaps the way forward is to reorient how sustainable development affirms dignity. Dignity is first and foremost an autonomous social relation constituting a demand. The demand is to break apart systems of exploitation and oppression, to "actualize the utopian potential immanent to a historical situation," as Žižek has poignantly stated.[46] The structure of the demand is social responsiveness that appears when different social and environmental struggles connect and inform one another. The outcome is a social bond premised upon the exercise of radical freedom. The demand for dignity carries a transformative potential because it institutes change by positing new spaces and times pregnant with potential. In this way, the demand for dignity is a calling into existence through an act of commoning. Dignity thus functions relationally. It is through the demand for dignity that the work of dignity is called into existence. The relational and transformative form of dignity can never be isolated in the single being of human existence. Rather, it relies upon broader environmental and cross-generational connections that belie narrative descriptions of what it means to be a human— or moral agent, for that matter.

Dignity is not self-sufficient; it is contingent upon spatial and temporal circumstances that are resistant to the insulated category of an inalienable right. We can see with the ongoing philosophical debate over the meaning of dignity that fine-tuning and clarifying its definition do not necessarily address why dignity matters and what exactly is at stake with the different meanings ascribed to the concept. What is important is not so much having dignity, where the emphasis is on a moralizing or legalistic view of dignity as a property—both of which are liberal constructs premised upon individualism and property rights. Put differently, it matters how dignity works.

Both development and design in the public interest are epistemic practices. They are ways of knowing, representing, and interpreting subjects. As such, they are both practices of subjectivation. As it is used in development, however, dignity emerges as an immutable and therefore absolute reality. It offers a stable platform from which to make sense of difference and from which to assess the value of a given project, functioning as a moral and legal corrective to everyday life. Here dignity becomes an essentializing concept, turning the body into an object of regulation and standardization. But the cultural artifacts this concept of dignity is part of constructing—the highrise housing or the pay to use water sanitation facilities—are merely symbolic representations of dignity. But the symbolic reality of dignity that such development produces is not the whole picture. There also exist uninterrogated excesses that constitute the symbolic power of the cultural artifacts of dignity. Failing to take into account this excess is where the domesticity of development lies.

Engaging with the structure of dignity, and how this works as a social relation, is an important step to alleviating the multifaceted ways in which poverty, racism, environmental degradation, and gender bias operate. Promoting the dignity of poor black bodies by framing development through the lens of human rights, the solution of which is to introduce new market opportunities that are built around the provision of much-needed water and sanitation services or housing, dissolves an otherwise political issue of emancipation into a moralizing problem of individual self-respect and self-worth. Freedom is a political problem of how dignity works to liberate all life from the oppressions of environmental degradation and exploitative socioeconomic structures.

So allow me to conclude by returning to the opening anecdote: I love you; I hate you! This is the mouthpiece of anticolonial struggle, when dignity defies the demoralizing landscape of poverty, pollution, and privation. This moment goes way back in time. Standing there at the side of the road as sewage-filled water jet-black from toxic chemicals runs between us, it strikes me that no matter how many sundowns of colonial violence have passed, or the guilt-ridden gestures of "first world" love have been made, or no amount of "first world" bills filling the bribery coffers of opportunism can ever reconcile, the love that a history of hate owes is complicated, messy, and rarely a two-way street. It can never be absolved in exchange for dignity as tyrannical monarchists, like the World Bank, or parasitical corporate aristocrats hope to do.[47] This hard yet hopeful realization is where many forms of international development miss the mark. For here in the slum, environmental degradation has compelled capital to reorganize, exposing the embryonic agency of commoning (collectives of people, animals, skills, energy, money, time, and resources). Dignity is the touchstone of such agency.

That antagonist contradictory logic at the core of "I love you; I hate you!" is the most candid way of describing the dynamic of capital consuming, producing, and reproducing life in the contemporary world. Enjoyment grafting onto excess, desire sustains the movement of excess as it actualizes in the commodification and privatization of dignity.

The work of dignity is to access and disclose the revolutionary power of sociality. Working dialectically dignity tensions love against hate, bringing hate to its knees with a leap of faith in a world that could be otherwise. Another world to that of the violence of war, exploitation, oppression, and the continual manufacturing of lack oiling the wheels of a global neoliberal market economy.

7

URBAN CLEAR-CUTTING

The greatest products of architecture are not so much individ-
ual as they are social works; rather the children of nations in
labor than the inspired efforts of men of genius; the legacy of
a race; the accumulated wealth of centuries; the residuum of
the successive evaporations of human society—in a word, a
species of formation.

—Victor Hugo, *The Hunchback of Notre Dame*[1]

Looking across the sand to where the Red Sea had lapped a few decades ago, I felt overwhelmed by the midday heat burning across the desert of the Jordan Valley. I stood in the silent ruins of what had been a fashionable restaurant on the West Bank. The rich aromas characteristic of Middle Eastern cuisine had vanished. All that remained of that once-popular tourist venue was a bullet-ridden, crumbling structure.

Eyes squinting from the sun's reflection across the sand, Ibrahim reminisced. He shared with me stories of a building filled with the carefree laughter of vacationers sipping cocktails and feasting on Palestinian delicacies. They had been captivated by the beauty of a desert sky at sunset and seduced by the flamboyant festivities that took place there. This had been a building where people looked forward to tomorrow. Buildings can do that. They can galvanize the future within the present moment, or they can squash the life out of it.

Today, like so many buildings in the Palestinian territories, all that remains are structures in mourning, buildings that threaten anyone who dares venture there with the memory of an ongoing conflict over land rights, resources, and incompatible versions of history.

Environments littered with the leftovers of conflict clip the wings of hope as the uplifting and dependable experience of Heideggerian dwelling shifts to vulnerability and misery. Conflict desecrates the friendliness of the built environment. It flattens the heroic monumentality of the built landscape to a horizontal memorial of collateral damage.

When the environmental movement intersects with urbanism, the outcome is typically sustainable urbanism and development and green design.[2] The desecration of the built environment because of warfare is not an issue environmentalists typically seize upon. In sustainable urbanism greening the city initiatives work with the presupposition of a peaceful city.[3] There is nothing wrong with this, only that it eclipses the environmental challenges urban war zones present. With the number of countries completely free of internal or external conflict falling to just ten in 2016, the environmental challenge conflict poses for urban living is increasing in importance.[4]

To the extent that environmental urbanism is confined to a nonviolent social condition despite millions of people being displaced by war, the weight liberal political thinking carries in the realm of environmental politics is exposed.[5] Liberalism is anchored in a nondisruptive conception of political life—consensus and incremental change.[6] Indeed, as I argue in the opening chapters of this book, incremental change often cloaks the co-optation of the revolutionary potential of environmental politics. And so I return to a central theme of this book: environmental politics cannot shirk from the challenge structural violence presents to all life on earth it if is going to be a transformative politics. This is not to say that politics is directed toward some indistinguishable system. That would mark the end of political action. But it does mean that politics begins in fidelity to the outcast and exiled, the victims of violence, and the events that interrupt how violence functions in the twenty-first century.[7]

My argument here is that if environmentalism neglects to engage with the violent terrain on which politics is conducted, revolutionary politics doesn't acquire any substantive traction in the movement, revolutionary in the sense that such political practices or theories aspire to end structural violence. A consideration of "cities under siege," as Stephen Graham has described it, is not just an environmental topic but also the basis for widening the frame of environmental politics and potentially part of recasting the role environmentalism plays in geopolitical practices and discourse.

UNIFORM REMOVAL OF URBAN LIFE

Prior to industrialization the majority of the human population lived in rural areas, relying on hunting, fishing, and agriculture to survive. In 1800 the urban population was a mere 2 percent of the global population.[8] Today, more than half of humanity lives in towns or cities, with the global urban population projected to increase to 2.5 billion by 2050.[9]

The world has urbanized rapidly since industrialization, and with it has come slum growth and mounting housing, transportation, infrastructure, and energy challenges. I examined the issue slum growth poses for environmentalism and development in the previous chapter. Here I want to focus on the importance urban living has for the majority of people on earth. Urban environments are the primary human habitat and will become home to many more people in years to come. In short, the human species is principally urban.

When considering what constitutes an urban population, there are a variety of measures that are used, such as population density, the size of a city, land-use patterns, and travel time to a large urban center.[10] The U.S. Census Bureau defines an urban population as an incorporated place with a minimum of twenty-five hundred residents.[11] The European Union (EU) adopts a regional approach in its classification of urban and rural populations, preferring to describe areas as predominantly rural, intermediate, or urban. Using this typology, a predominantly urban population is one where the "rural population in rural grid cells accounts for less than 20% of the total population."[12] If at least 50 percent of the population lives in an urban center, the EU describes this as a city. A town or suburb is one where "less than 50% lives in an urban center but more than 50% of the population lives in an urban cluster."[13] The Indian census classifies an urban population as one with a minimum population of fifty thousand, where 75 percent of males are not engaged in agricultural work, with a density of at least four hundred people per square kilometer.[14] The urban population continues to grow as the global population increases and more people migrate to cities from conflict zones and in search of economic opportunities.

Cities, however, are not only a physical space that can be measured according to their size, volume, and shape.[15] Edward Glaeser has characterized cities as the "absence of physical space between people and companies. They are proximity, density, closeness. They enable us to work and play together, and their success depends on the demand for physical connection."[16] Cities provide a platform for human interaction and a place to reap the benefits of enhanced social relations. Living in close proximity can bring enormous benefits, such as conserving important resources and energy along with sharing a variety of social and cultural services providing access to health, education, entertainment, and transportation.

Moving against the grain of history and with an unapologetic nod to the highly acclaimed book *The Death and Life of Great American Cities* by Jane Jacobs, Leo Hollis has rejected the idea that cities are bad for people.[17] Indeed, he mounts a very convincing case as to why cities are good for us.[18] "The city is a place where strangers come together," Hollis writes, and it "is possible to think that the metropolis is perhaps our greatest achievement."[19] For him, the city is at the center of addressing the many problems humanity currently faces. The challenges presented by "climate change, unprecedented migration, the depletion of natural resources and a widely perceived decline in the civic values that hold our societies together" have positioned the human race between "disaster and survival."[20] Hollis maintains that urban living is our best bet for survival. In order to maximize the positive potential of urban living, cities need to be engines for social equality, democratic participation, hubs of creative entrepreneurialism, vibrant public life, and, most of all, trust.

In order for this to happen, a lot has to change. Livable cities need good planning and design, as well as inclusive policies that create more opportunities for human flourishing and positive social interactions. Environmentalists and members of the sustainable design community have for a long time pointed out the importance of greening cities as a way to combat climate change and to improve quality of life (for both human and other-than-human animals). This can be done by designing energy-efficient buildings, incorporating clean-energy technologies into the built environment (solar panels and wind turbines), improving public transportation and services, outfitting the city with rooftop gardens, parks, bicycle paths, and stormwater capture systems, creating inexpensive housing or new habitats using native plants and flowers, and last but not least welcoming public spaces. This has led to the development of various green rating systems, certifications, organizations, and training initiatives such as LEED (Leadership in Energy and Environmental Design), SEED (Social Economic Environmental Design), GSBC (German Sustainable Building Council), and IGBC (Indian Green Building Council). In a nutshell, an eco-friendly city is more energy efficient, has optimum air and water quality, strong environmental governance, good waste management, affordable housing, is inclusive, and has extensive environmental infrastructure and eco-friendly services. War, violence, and persecution render all of this impractical.

Thus far our entrée into the twenty-first century has been rocked by wars on several fronts, wars that have left large parts of the world uninhabitable. The twentieth century closed with the Yugoslav Wars (1991–2001) still raging, resulting in cities like Mostar losing important cultural landmarks such as a sixteenth-century Ottoman stone bridge, which also cut the town off from its only source of potable water.[21] Since the beginning of the twenty-first century the Syrian city of Aleppo has all but vanished off the face of the earth as war has torn that country apart. Meanwhile, Gaza, the target of an Israeli-Palestinian political deadlock, continues to hold its Palestinian inhabitants hostage as they barely carve out an existence in the rubble they call home. The invasion of Afghanistan and Iraq by the "coalition of the willing" united in conducting a "War on Terror" (2001–2013) turned cities across both countries into hostile environments.[22] The Ukraine city of Pervomaisk is but one of several cities and towns barely standing because of fighting between government forces and rebel groups.[23] And then of course there are the many isolated attacks that include the attack on the Twin Towers in New York City and suicide car bombers in Baghdad.

Yes, the twenty-first century has kick-started with quite a bang, reducing entire human habitats to dust and debris. During wartime, bombs, bullets, and battle transform sturdy buildings into Swiss cheese and streets into craters. Infrastructure, such as electrical grids, hospitals, sewers, and water-supply systems are decimated. This is the incidental damage that comes with conducting war, or it is the result of noncombatant domains and civilians being deliberately targeted in an effort to bring a community to its knees. Regardless, when peace becomes a memory concealed somewhere beneath the wreckage civilization has left behind, cities hollowed out by violence leave people without places to live, work, and flourish.

In 2013 Noeleen Heyzer, head of the United Nations Economic and Social Commission for Asia and the Pacific, reported that more than 1.5 billion people live in unstable and conflict-affected areas.[24] By the close of 2015 65.3 million people had been displaced by war and violence. That is 5.8 million more than the year before.[25] In 2015 twenty-four people fled every minute because of war and persecution.[26] As of 2016 the Global Peace Index estimated the global economic impact of war at US$13.6 trillion, a total of 13.3 percent of the world's economic activity.[27] Between 2008 and 2016 the world became 2.44 percent less peaceful, and deaths because of terrorism increased 286 percent, while deaths as a result of war increased more than fivefold.[28]

In cities ravaged by war the material conditions of everyday life are seriously diminished. Whether we are referring to the collateral damage of conducting warfare or a conscious targeting of the built environment as a way to erase a community's identity and its capacity to survive, war systematically destroys an important human habitat. In many instances, such as Gaza and Aleppo, such cities could be classified as collapsed urban ecosystems. Their infrastructure, housing, food and water supplies, public institutional spaces such as hospitals and schools are either severely lacking or they quite simply don't exist anymore.

Given how dependent people are upon a well-functioning urban infrastructure, more and more its destruction has shifted from collateral damage to a means of waging war. Graham explains, "The very nature of the modern city—its reliance on dense webs of infrastructure, its density and anonymity, its dependence on imported water, food and energy—create the possibility of violence against it, and *through* it. Thus, the city is increasingly conceived of as the primary means of waging war by both state and non-state fighters alike."[29] Studying the air strikes of Syrian and Russian government forces, Amnesty International provided this disturbing assessment: "Syrian and Russian forces have been deliberately attacking health facilities [as] part of their military strategy."[30] In July 2015 the water supply for the capital city of Kosovo, Priština, was cut off over concerns that ISIS was going to poison the city's water supply as an act of war. More than two hundred thousand people were affected.[31]

Gaza has all but been gutted as a result of Israel's three military offensives (December 2008–January 2009; November 2012; July–August 2014).[32] The Palestinian Water Authority has estimated the cost of damage to Gaza's water and sanitation infrastructure to be in excess of $34 million.[33] The destruction of seventeen hospitals and fifty-six primary health-care facilities during the 2014 military offensive left 2.5 million tons of debris.[34] Not to forget, the 2014 operation also destroyed 18,000 housing units and damaged 44,300, 25 schools were destroyed and 122 damaged, 20 to 30 percent of the water and sewerage network was damaged, as was the water-desalination plant in Deir al-Balah and the electrical network.[35] Since 2016 de-development, a term that refers to the reversal of development, in Gaza as a result of three military offensives in the past six years and eight years of an economic embargo could potentially leave the remaining 1.8 million Gazans living in an uninhabitable environment by 2020.[36]

In short, the damage and destruction war inflicts on the built environment is nothing short of urban clear-cutting. Just as important wildlife habitats are decimated when a forest is clear-cut, with the sophisticated war machinery of the twenty-first century, in addition to suicide bombings, and the increasing use of improvised bombs such as barrel bombs, which wreak havoc because of their imprecision, we discover a similar kind of violence and erasure being inflicted upon another habitat: the human one.

Certainly the idea of an urban environment as akin to the rich ecological system characteristic of a natural forest may at first seem like a terrible stretch of the term, and given that clear-cutting is responsible for more than half of the world's rain forests disappearing, to some it may also seem unconscionable to describe the leveling of entire cities as a specifically urban instance of clear-cutting.[37] But war uniformly destroys a city in the same way that clear-cutting rain forests does, leaving the ground bare of buildings and infrastructure.

Clear-cutting forests meets the human demand for agricultural land, housing, and wood by dominating the complex ecology of a tropical forest and turning it into a monoculture system with low diversity. The process leaves behind very little regrowth, making it difficult to regenerate. Clear-cutting trees seriously interrupts and degrades the ecological and hydrological cycles of a forest as well as removing an important habitat on which wildlife and indigenous groups depend.

Clear-cutting a city is one way of seizing physical and psychological control of an area and its inhabitants. The physical damage incurred from clear-cutting urban areas exacerbates pollution as important waste infrastructure is destroyed, increases flooding as storm-water systems are wrecked, the environment is littered with mounds of debris filled with toxic materials, and quality of life is negatively impacted. Clear-cutting of both a forest and a city are applications of coercive power. Power is exercised over another in an effort to subjugate, not empower it. Success comes at a complete disregard for the diversity of that which is dominated.

After a city is clear-cut, what remains is so sparse that inevitably the city will need a huge investment of economic, social, cultural, and political will and resources to bounce back. In this way urban clear-cutting reduces life in the city to the pure present of actually existing circumstances. In effect, human experience shrunken to the sphere of actuality and the pure present limits human freedom.

COLLECTIVE PUNISHMENT

On the morning of August 2, 2014, Israeli Defense Forces bombed the northern city of Jabaliya in Gaza, directly targeting mosques they claimed were storing weapons. The ancient Al-Omari Mosque, dating back to 649, was destroyed. The mosque had been renovated after incurring damage from attacks that took place in 2008 and 2009. In addition to being a significant cultural heritage site in the Palestinian territories, the Great Mosque, as referred to by locals, was an important place of worship (a religious building that could house up to two thousand worshippers) and community gathering. Many other historic houses of worship—Omar Ibn Abd al-Aziz Mosque, Mahkamah Mosque, the Orthodox Church of Saint Porphyrius—were all shelled during the conflict.

Similar instances of cultural erasure abound in other conflict zones. On Sunday, August 23, 2015, ISIS detonated the ancient temple of Baalshamin in Palmyra. The temple was an artistic treasure dating back to the first century and was later expanded by Emperor Hadrian.[38] The fundamentalist group went on to destroy the ancient Aramaic city of Palmyra, where the temple had stood.[39] Also known as Venice in the Sands, Palmyra was an oasis in the desert situated between the city of Damascus and the Euphrates River. Merchants traveling to Damascus or Emesa from Babylon had once stopped there on their way. During the first century, the city came under the control of the Romans and was transformed into an important trade route connecting the Roman Empire to Persia, India, and China. Unsurprisingly, then, the architectural treasure of Palmyra incorporated a magnificent admixture of Greco-Roman and Persian styles. These architectural styles went on to influence the revival of Western classical architecture and urban design after travelers during the seventeenth and eighteenth centuries exported the aesthetic discoveries they made in Palmyra to the West.[40]

Oh, the contradiction of a highly disciplined military operation fanatically faithful to a socially rigid and hierarchical view of the world slipping into monstrous revelry at the destruction of ancient heritage sites. Taking pickaxes, sledgehammers, bulldozers, and dynamite, they destroyed with glee and great fanfare the traces of pre-Islamic or Christian heritage. Completely driven by a fundamentalist passion, one that is in stark contrast to the political correctness of the secular West, ISIS expands its network of terror using the spectacle of Western media to terrorize those who participate in the event as voyeurs. Using the fascination of the spectacle of violence against itself, both ISIS and the Western voyeur participate in the fascistic absolute together.[41] This may be why Alain Badiou has written, "All access to the real, all real certainty, involves the mediation of the infinite. Thus, there is an organic relationship between the infinity of being, the access to this infinity of being, the various different truth procedures, and the fact that the subject is the instigator of all this, inasmuch as it summons the individual to this action or process."[42] Coming into contact with the absolute incorporates the collective into a political subject, and this is a subjective process.

By the close of 2015 ISIS had looted and damaged many more ancient heritage sites, including the Iraqi cities of Mosul, Hatra, Ninevah, Khorsabad, Nimrud, and the Syrian cities of Apamea and Al-Qaryatain.[43] If the "city is a repository of history," as Aldo Rossi noted, then ISIS deliberately refashions the built environment according to its version of Muslim history. It does so by a system of elimination: removing traces of histories that fall outside the parameters of ISIS's distorted version of

Islam.[44] This is an attempt to both give fundamentalist belief a form and close history off from the "infinity of being."[45] The irony is that it reintroduces the dialectic of historical change in its efforts to eliminate it.

Then we come to the city of Aleppo, a UNESCO World Heritage Site. It has incurred some of the worst damage during the Syrian civil conflict. The Great Mosque of Aleppo, dating back to the Umayyad period, the city's ancient market of Souq Al-Madina, not to mention Roman streets and homes from the Ottoman period, have all been damaged if not completely razed during the fighting. Irina Bokova, UNESCO director-general, issued an urgent plea to all involved in the conflict "to immediately stop the destruction of the country's cultural heritage, and in particular Aleppo. Heritage should not be taken hostage in the conflict and I condemn any military use of cultural sites, as targets or as shelters . . . Damage to the cultural property of the Syrian people is damage to the cultural heritage of all humanity."[46] Bokova has drawn attention to the city as a "material artifact," as Rossi described one view of the city.[47] It is a text that changes over time, and along with each and every addition and modification made to it, a repository forms. As such, the city is a living document that offers valuable evidence concerning human beliefs, values, rituals, ways of life, aesthetic tastes, and intellectual pursuits and how these change over time.

Rossi insisted that a city is not only a library storing important information about the past but also can also tell us a great deal about the process of history itself. The various historical layers and the relationships between them cast a spotlight on history in the making. This is the "idea that the city is a synthesis of a series of values" concerning "the collective imagination" that "represent ideas of the city that extend beyond their physical form, beyond their permanence," Rossi explained.[48]

Portrayed in this way, the destruction of cities is part of a longer trajectory of the history of civilization. The idea of a city persists regardless of whether or not the city is physically present, "thus we can also speak in this way of cities like Babylon which have all but physically disappeared."[49] If this is the case, then the appeal to conserve important heritage sites for their intrinsic universal value is all but moot. Furthermore, according to this view, collective memory is not a creative process that changes over time. Instead, the assumption is that collective memory is fixed. It is an entity that can be handed down across time. Additionally, the problem of conserving the idea of a city becomes one of how it can be imparted to a subject, as though the subject in question is also a given. This approach does not address the political issue of how a subject is constituted by the destruction of the built environment.

Irrespective of being collateral damage or a conscious act of targeting important infrastructure and cultural heritage sites, central to a wrecked city is the event of ruin itself. Every such event produces a subject. So what becomes of a subject constituted through the unforgiving landscape of war and persecution? Living in a built environment petrified by violence, one where its use value (infrastructure and housing) and collective memories have been disemboweled, what remains of human experience?

As of 2016 there are entire generations who have grown up in a conflict zone. They have not attended school, and if they have, it has been irregular at best. People are living under constant threat and bombardment, on the brink of death and starvation, and where not even schools or places of worship can ensure safe refuge anymore. Entire communities have been consumed by fear and forced to live underground to escape the turmoil. People have lost trust in their governments, international diplomacy efforts, and nonprofit organizations to ensure safe passage out of the city and within it. In the war zones of the twenty-first century reports fill the media of people being arbitrarily detained, beaten, tortured, abducted, arrested, and starved, while others quite simply disappear or drown trying to escape in leaky boats off the Mediterranean coast.

They carry with them a formal idea of their city, an idea that is constantly in formation; the idea is therefore a means through which to give form to the events taking place. This Idea is both retrospective and an action. Speaking of formalization and the communist Idea, Badiou has outlined the two-fold sense of such an Idea "owing to the fact that it makes possible to identify the general form of the current political movement, and in the sense that it's also normative," that is to say, "it allows you to judge or valorize certain situations over other tendencies."[50]

The genocidal practice of ethnically cleansing a population by targeting its cultural heritage breaks apart the connections that tie people to a place, to one another, and to their past. The erasure of the built environment inevitably extinguishes how people spatially and temporally orient themselves in the world. And living under arbitrary conditions means living in a state of high alert. It doesn't just strip people of their past but also interrupts and distorts their relationship to futurity. An environment constituted by indiscriminate events whose effects are punitive is an environment where human rights cannot flourish. Considered in this way, one of the greatest human rights challenges war presents is environmental: living in an inhospitable habitat.

With urban clear-cutting the aged skin and bone of a city evaporate. Time infuses a sensuous textuality throughout the spatial layers of a city. Different temporalities constitute the quality and character of the built environment. They unleash an admixture of spatialities that are brought to life through different materialities and bodies, formal organizations, the interaction of volumes and shapes, contrasting outlines, shadows and lighting, flora and fauna, and the rituals people engage in. Additionally, there are the smells, sights, sounds, and cadences inhabiting every nook and cranny of a city. Feeling at home in a place is a synthesis of all these elements. These contract and form habitual ways of living, but they also contain in their combination the potential for new experiences that include in their purview a reinvention of the past.

When the past is eliminated and the tabula rasa of a detonated ground is infused with exhausted bereavement and a minefield becomes home, life quickly narrows to an endless repetition of the same: the same pure present repeats over and over again ad nauseum. War confiscates the imaginative layering of history, memory, and experience, and most of all the "creative evolution" of hope in action, to use a conceptual pairing Henri Bergson coined.[51] In the war-torn worlds of Aleppo, Gaza, Baghdad, and Pervomaisk, belonging gives way to warrior subjectivities— fighting to remain, fighting to survive, fighting against all odds, fighting to become immune to their environment.

The differences that constitute urban living are univocal; they cannot be subsumed under a unified identity. Hence, when I speak of urban clear-cutting erasing belonging, I am not referring to an experience a person owns, as though that person were a unified entity that remains constant over time. Rather, belonging is understood as a condition without identity; it is not the result of processes integrating the many differences constituting urban life. Nourished by memory, the calls of history, and the trivialities of habitual living, belonging facilitates day-dreaming and emancipatory imagining, without which hope sours and turns to revenge.

Conflict creates environments of entrapment and separation. Sure, there are random moments of flourishing, happiness, and optimism despite the destruction and violence permeating a community at war: short, brisk occasions where children turn piles of bricks into an obstacle course, where the overexposed insides of a ruin are transformed into makeshift soccer fields, or collapsing walls provide a canvas for painting and drawing, or even small areas cleared of debris where the remaining civilians grow a few vegetables or collect rainwater.

Architect Eyal Weizman defiantly recomposed an abandoned Israeli military post in the occupied territory of the West Bank into a bird-watching facility for local Palestinians, transforming the exclusive and hostile function of the architecture into an inclusive and welcoming one. His detailed maps of Israeli settlements in the West Bank don't just "mark the settlements as mere points but describes the actual form of their layout, shows that, beyond the mere presence of Israeli settlements on occupied land, it is the way they have been positioned, designed and laid out that directly and negatively affects the lives and livelihood of Palestinians."[52] His practice is committed to challenging how architecture and planning are used as a mechanism of state control. Weizman demonstrates how the physical transformation of Israel-Palestine is part of Israeli military strategy and a larger system of power and control. This places Israeli architects and planners in a difficult situation, where the whole question of artistic integrity and responsibility inevitably comes up. Israeli architecture and planning are implicated in the gradual erasure of Palestinian history, culture, society, and economic life. In short, Weizman's work is an environmental intervention into the structural violence embedded in spatial arrangements.

Throughout human history many cities have been rebuilt once war had come to a close. Rotterdam, Dresden, and Beirut are just a few examples. So the question of the war-torn city as an environmental challenge is less a problem of how to rebuild in the aftermath of war than one of why the targeting and subsequent leveling of the built environment are a specifically environmental problem.

Usually environmental issues are relegated to that nebulous sphere called nature, and environmental justice issues examine the impact environmental degradation has on human health. But if we can all agree that human activities have radically changed, modified, and shaped the natural world, then the sharp distinction formed between nature and human society rapidly fades. Instead of trying to split hairs over what nature *is*, the premise of this chapter has rested upon the question of how nature *works*. Nature is culturally constituted and socially organized. It is the by-product of how human beings imagine and create the world. Yet it also exceeds human authorship.

As masses of "vibrant matter," to borrow from Jane Bennett, cities are self-generating and differentiating.[53] Their liveliness is constituted through the open interconnection of materiality, bodies, energy, and forces. In their combination these take on different kinds of investment. They can be liberatory, as in the case of Weizman's practice, or oppressive, as is the case with ISIS. This shift in emphasis allows us to expand our understanding of society as narrowly confined to the realm of human institutions, organizations, and communities to include ecological processes and other-than-human species. It also enables us to think of urbanity beyond the dominant frame of growth, density, and prosperity.

Destroying the built environment, an environment human beings collectively create over long periods and that they come to depend upon for their survival and through which a rich cultural heritage is expressed, is an environmental issue. Just as a conservationist might argue for the protection of the wilderness on the grounds that it is a natural beauty as well as home to any number of animal species, so too the destruction of the built environment annihilates the cultural legacies of human civilization, along with spoiling one of the most important of human habitats: the city.

If the city is a place of "community, complexity, and creativity . . . a space that has more in common with natural organisms such as beehives or ant colonies," as Hollis has put it, then just as in any ecosystem, there is a food chain.[54] In a globalized world this runs from South to North. As Graham has aptly pointed out, too often urban studies "systematically ignore the way the North's global cities often act as economic or ecological parasites, preying on the South, violently appropriating energy, water, land and mineral resources, relying on exploitative labour conditions in offshore manufacturing, driving damaging processes of climate change, and generating an often highly damaging flow of tourism and waste."[55] The urbicidal violence Graham identifies facilitates capital accumulation in the wealthy cities of the global North at the expense of the global South. The urbicidal violence of urban clear-cutting highlights another structure of violence, one that is not entirely unrelated to what Graham identifies and yet is distinctly different.

The leveling of cities and the disappearance of humanity's shared cultural heritage attest to an invisible structure of violence permeating life in the twenty-first century. They are testimony to a transnational society divided by wealth, power, opportunity, and historical privilege. So while we might look on in horror at the spectacle of war, it is important to remember that each and every one of us is implicated in the violence being witnessed from afar. In the words of the former UN secretary-general Ban Ki Moon, "We are facing the biggest refugee and displacement crisis of our time. Above all this is not just a crisis of numbers; it is a crisis of solidarity."[56] And solidarity in this instance must begin from the standpoint of those who are excluded.

8

PROTEST WITHOUT PEOPLE

Have you ever noticed how slow and silent the world becomes after disaster strikes? I was in Paris during the bombing and shootings that killed 128 people (November 13, 2015). The city entered a state of shock, and all public life came to a dramatic standstill. I recall catching an eerily empty metro the next morning and walking the vacant streets in the Fifteenth Arrondissement, where I was briefly living. The bustling sounds of the city had ground to a halt. Even the tone of police sirens dampened. The sense of solitude and stillness that arose alongside a global outpouring of solidarity with Paris was juxtaposed by the frenetic spectacle of terrorism spinning through the news media.

Media headlines responding to the Paris attacks were thunderous. The front pages of the world's newspapers declared, "The Horror" (*L'horreur*—*L'Équipe*); "The Bloody Siege of Paris" (*Daily Telegraph*), "Paris Terror Attacks Kill More Than 100"; "Paris Declares State of Emergency" (*New York Times*); "The War on Paris" (*La guerre en plein Paris*—*Le Figaro*); "Massacre on the Streets of Paris" (*Independent*); "Carnage in Paris" (*Carnages à Paris*—*Libération*); "Massacre in Paris" (*Sun*).[1] The cumulative effect of the media frenzy in the West was a shared imagination filled to the brink with an unthinkable excess that suddenly seemed to lurk in every nook and cranny of Parisian life.

Afterward the Western political imagination rapidly escaped into apocalyptic scenarios that envisaged Western values under siege by Islamic fundamentalists. This was a potent collective imagination, so persuasive and charismatic that it ripped apart the *common* condition underlying political life. With the *common* of politics betrayed, the road was cleared for political authorities to prohibit with minimal backlash public demonstrations scheduled to occur around one of the world's most internationally important events for even the most mainstream of environmentalists—the United Nations Framework Convention on Climate Change.

Through apocalyptic imagination, catastrophe loses its clutch on finitude and becomes mythic. As a result, the emancipatory promise of political imagining becomes timid as it is cast adrift in the raging tangle of "disaster." Nonetheless, all is not lost. By virtue of its capacity to loosen the mythic grip that apocalypse holds over the imagination, imagining harbors emancipatory inclinations.

Political imagining, as I understand it, is a constellation of four ideas. First, imagining is a mode of distributing "our shared sensible world" in the way that Jacques Rancière has defined the politics of aesthetics.[2] Second, it is a performative operation that, following Judith Butler, brings ingrained norms into conversation with the aberrant, such that the aberrant either functions as the point of reference that reinstates normativity or it creates a site on the margins of normativity where emancipatory cultural and economic politics influence social life.[3] Third, it is a "de-forming agent" in the way that Walter Benjamin described imagination, always referring to "something formed beyond itself."[4] Finally, political imagining employs the logic of a promise as articulated by Hannah Arendt.[5] When these combine in an emancipatory operation of imagining, they forecast what Braidotti had called a posthumanist future.

The argument developed here is that the emancipatory or repressive exercise of imagining depends upon whether or not it combines an outward-looking position that invites other realities to expand and deepen our own with a critical inwardness; or whether it pulls up the barriers around an inward orientation, where one feels responsible for and committed only to their own reality.

Moving forward, imagination is one of the biggest challenges environmental politics faces. We know that the world's environment has been and continues to be seriously harmed. What we don't know is the full impact of this crisis, because it all depends on what we do about it from here on out. The biggest constraint we face vis-à-vis our own future is the hard-nosed reality that the human species is killing itself. Consequently, the current generation not only faces a remarkable opportunity to make a real difference in the world, but also carries a heavy responsibility to look beyond the current socioeconomic paradigm that produces widespread suffering and harm. Indeed, as previous chapters have shown, the success of the former relies upon a commitment to the latter. The question that now presents itself is, how do we tackle this? In large part the answer lies with the imagination.

APOCALYPTIC IMAGINATION

His finger on the pulse, Fredric Jameson has declared, "It is easier to imagine the end of the world than to imagine the end of capitalism."[6] More recently, cautioning against the "normalization of a catastrophic imagination," Brad Evans and Henry Giroux have pointed out that "under neoliberalism, imagining a better future is limited entirely to imagining the privatization of the entire world or, even worse, imagining simply how to survive."[7] The brutal effect of this situation is that violence has become normalized.

Normalization works in mysterious ways. On the one hand it is habitual, on the other hand, as Butler has taught us, it is a performative operation. Normalcy carries an other, it's shadow, an exteriority upon which it depends: "The normative force of performativity—its power to establish what qualifies as 'being'— works not only through reiteration, but through exclusion as well . . . those exclusions haunt signification as its abject borders or as that which is strictly foreclosed: the unlivable, the nonnarrativizable, the traumatic."[8] Indeed, what makes the norm normative is the performance of the other as other.[9] In this respect the normalization of violence is misleading because it always includes in its habitual constitution an aberrant force. As environmentalism increases in popularity, it concomitantly performs a deviant environmentalism, as discussed in chapter 4.

Nightmarish scenarios appear as the mainstream grapples with reconciling populist and deviant forms of environmentalism. Some examples: the world's misfortunes come from aliens occupying earth; comets the size of New York threatening to crash into the earth and obliterate the human species; floods biblical in scale and monstrously foreign in their power submerging the earth; contagion cruelly invading human bodies; ferocious wildfires bringing life to the breaking point. All present a world terrorized by natural disasters, but this is also one where the structural violence of capitalism remains innocent. These Hollywood movie scenarios propound the false belief in an

environmental crisis unrelated to the system of global capitalism. They isolate the suffering and impoverishment capitalism
produces from the history of inequity and the relentless exploitation of humans, other-than-human species, and the environment as a whole.

One of the biggest hurdles the environmental movement
faces is not a lack of imagination. With the rise of mass media
and political spectacles, there is no absence of imagination; rather,
the question is how imagination under late capitalism works.
If an "environmentalism of the rich," as Peter Dauvergne has
fittingly called it, is to be broken to make way for a more transformative politics, then the environmental movement urgently
has to find a way to reignite the emancipatory potential of
imagining.[10]

The content of an apocalyptic imagination exists in spaces we
don't inhabit. Hence, these spaces become larger than life. In
Dante's Peak (1997) audiences are gripped by the hypnotic horror
of volcanic ash blackening the sky and lava heating up the earth.
Featuring a celebrity lineup that includes Kate Winslet, Matt
Damon, Jude Law, and Gwyneth Paltrow, Contagion (2011)
presents a perilous world overtaken by a virus that is rapidly
infecting and killing off the population. Cruel waves of heat
threaten to leave the world breathless and thirsty in the International Emmy Award winning Australian television movie
Scorched (2008). Similarly, in Backdraft (1991), the second-highest-
grossing film about firefighters, intense fires destroy the built
environment, killing, disfiguring, and injuring the human
body.[11] In the epic movie Noah (2015) God punishes human
beings for their sins by letting loose a flood of biblical propor-
tions, in the way only a transcendent force can, drowning nearly
every single human being on earth, bar one family and a slew of

beasts. Moving in tandem with the dramatic and haunting soundscape by Clint Mansell, the vengeful waters of God's wrath submerge the earth under a blanket of darkness and the adversarial forces of nature in the raw.

It is not only the movie industry riding the wave of natural disasters and their currency in the popular imaginary. The format of "breaking news" sometimes even suspends regular programming to cover catastrophic events, such as Hurricane Katrina in 2005. The live broadcast of "disaster marathons," as Tamar Liebes called them, ritualize the emotive charge of violence, turning the spectator into a witness.[12] It is important to acknowledge that news coverage following a natural disaster is a necessary public service providing much-needed information. Robert Putnam has called this prosocial programming, the effect of which is often altruism as people are prompted to donate either their money or time to relief efforts.[13]

The kind of programming Putnam is describing, however, is community television news such as PBS and the *PBS NewsHour*, neither of which sensationalizes the content of its stories. In these instances, the content is contextualized through informed discussion and thorough research, which accounts for the active listening and viewing Putnam identifies. Too often, however, the same images of destruction and loss are embellished by sloganistic journalism replaying the same images ad nauseum for their shock value, fragmenting the story through the use of punch lines and abridging history if mentioning it at all. All these devices take the heart out of a story, emptying it of nuance in meaning, contextual color, and the texture that comes from combining different perspectives. "To deny what was" Rancière has noted, "you don't even need to suppress many of the facts; you only need to remove the link that connects them and constitutes them as a story."[14]

The "power" of the spectacle is less about ideology as an instance of false consciousness, as Guy Debord maintained, than a matter of desire. The spectacle is an instrument of stimulation. It concerns the investment of social energies and forces in a simplified representation of history. It is far easier for people to digest a summary of otherwise terrifying realities than to confront the structural violence, which Slavoj Žižek had candidly called "objective violence . . . the violence inherent to [the] 'normal,' peaceful state of things."[15] That is to say, the messiness of inequity and injustice underwriting the quotidian.

Spectacle guards the mass's "wish for sleep," Debord insisted.[16] But are we sleeping? A Gallup poll published March 16, 2016, reported that concern in the United States about global warming was at an eight-year high, with 64 percent of Americans reporting they were worried about climate change.[17] After the 2013 and 2014 winter flooding in the United Kingdom public opinion on the human causes of climate change rose to the highest level since 2005.[18] And yet globally changing concerns over climate change align with which nations benefit the most from fossil-fuel emissions. An international survey conducted by the Pew Research Center in 2015 found that people who live in the two countries with the highest carbon dioxide emissions, the United States and China, are least concerned about climate change. Meanwhile, Latin Americans and sub-Saharan Africans are the most worried.[19]

We are fully aware of the environmental crisis we currently face. Watching it unfold on the screen or throughout social media doesn't actually change anything. Regardless, audiences flock to the movie theaters eager to enjoy the latest natural disaster flick. The success of the spectacle of natural disaster comes from viewers' actively participating in the energetic and hypnotic force of the spectacle, where contradictory emotions such as fear and fascination combine with the full realization that the journey is one of mere enjoyment freed of any kind of responsibility toward its subject matter.

In all the films mentioned the formula is simple: a messianic character (typically a white male) filled with courage and conviction emerges out of the rubble of annihilation. His secular heroism is firmly grounded in concrete actions that exist in tandem with his role as the ultimate savior. Put succinctly, the apocalyptic imagination operates as divine feeling by tensioning the tangible with the absolute.

Politics reliant upon an Absolute, as Gilles Deleuze under the influence of Benedict de Spinoza might have said, subordinates the nomadic and anarchical potential of a variety of political experiences to identity.[20] This is authoritarian, because the differential power (*puissance*) of political imagination is colonized by unidirectional affects, affecting but never itself being affected by the changes that structure reality. Under these circumstances political imagining can only ever function representationally by what Deleuze might have called a model of recognition colonizing the open and inclusive energies of the social field with a dogmatic image that is apocalyptic in both scale and affect.[21]

If we return to Jameson's astute observation of how easy it has become to imagine the end of the world, his dictum puts into question how imagination under global capitalism works. To add to Jameson on this point, I would go so far as to say that under global capitalism political imagination is increasingly taking apocalyptic form. The lucrative movie industry is built around the irrational desires imagination feeds off. Set against the backdrop of a very graphic end of the world scenario, *Noah* (2015) grossed $359.2 million worldwide. The fascination roused by storms of quasi-cosmic magnitude in *The Perfect Storm* (2000) helped bring in $329 million worldwide at the box office. Massive storms were the primary narrative device in *Cast Away* (2000), which grossed nearly $430 million worldwide. The special effects for which *Twister* (1996) is so well known plunge viewers into a close-up, albeit virtual, experience of the tornado's force, which is both exciting and terrifying at the same time. Unsurprisingly, the worldwide gross for *Twister* came in at $494.5 million. Grossing $544.3 million worldwide, *The Day After Tomorrow* (2004) was built around an extreme climate change scenario that abruptly moves the world from normal to disastrous.[22]

These statistics testify to the reflexive power of the spectacle. Think about it: why else have these natural disaster movies become worldwide box office hits if not for the *jouissance* spectacle incites. In them catastrophe is activated by a range of aesthetic devices that include special effects, theatrical soundscapes, immersive cinematic experiences, and the captivating star power of popular culture icons. It is a typical example of the atrophy of imagination under late capitalism that Jameson has identified.

Excited, aroused, horrified, intrigued, and entertained by masculine displays of strength pitted against the transcendent forces of nature, audiences the world over are attracted to the images and stories of disaster movies. Yet the power of an apocalyptic imagination is not restricted to the imagery of disaster as it appears in news media, social media, and the movies. More recently, it has become a lethal weapon of war.

The success of ISIS is standardly attributed to its very effective propaganda machine, a "sophisticated marketing and social media campaign."[23] However, it is a specious argument to accuse ISIS of ideologically influencing young people. The opposite is true. The myth of ISIS is viable only because of the continual public condemnation of ISIS. The very act of debunking ISIS masks the reality of populist fascination with it. This is precisely the stuff myth is made of and upon which ideology functions. To appropriate Žižek's brilliant inversion of Marx's definition of ideology—"Sie wissen das nicht, aber sie tun es" (They do not know it, but they are doing it)—"they know it, but they are doing it anyway"—the West has become fascinated with ISIS despite not believing in it.[24] And ISIS artfully employs the ideological operation Žižek describes.

While it is a no-brainer to deplore the threat ISIS made to blow up Syria's Tabqa Dam, which, if successful, would have released vast amounts of water onto Mosul and Baghdad, inevitably drowning out all signs of life, it is nonetheless effective. ISIS puts the affective power of imagining to punitive and authoritarian uses. Its political experiences are conditioned by a transcendental image of religiosity, whereby political identity is constituted through the realization of religious values and the transcendent and absolute power (*pouvoir*) of God.

The spectacle of environmental damage ISIS taunts the West with tensions the irrational machinations of unconscious activity with conscious rationalization. In this way it taps into the dark underbelly of the spectacle, taking people to places they would never typically go, not consciously, that is. Here the threat of environmental catastrophe exploits the most apocalyptic of Hollywood environmental scenarios, only this time fiction becomes potentially real. Indeed, the apocalyptic political imagination of ISIS consists of fantasy threatening to make reality.

The spectacle is a commodity fetish, and as such it enables the conscious negation of the harsh reality that human beings are destroying the very habitat they depend upon for survival. ISIS is a master of how to put the social energies and forces

driving commodity fetishism to work to advance its own political agenda, all the while denouncing the West and all it represents, including commodity fetishism. As such it is irrelevant
what we think about ISIS's use of spectacle; one has to admit it
does it really well. The threat to use natural resources as a lethal
weapon of war takes the spectacle to a whole new register.

The "imagination de-forms," Benjamin insisted, "it never
destroys."[25] This, for him, is what distinguishes the imagination
from fantasy. Fantasy, he explained, dissolves form. Fantasy
posits something that cannot be actualized. Imagination, on the
other hand, Benjamin said, is directed "at the ideas and at nature
in the process of de-forming itself."[26] In this way the imagination
is never entirely independent of reality. It dynamically combines
matter, sense, intuition, memory, and thinking.

In the case of apocalyptic imagination, the de-forming Benjamin accorded the imagination is one of misidentification. It
mirrors all the fears and anxieties of the social field, bringing
them together into a single coherent whole into which anxieties
are reconciled through the experience of pleasure. In this
regard an apocalyptic imagination functions like the Lacanian
imaginary.

Simply put, Jacques Lacan described that when a child ini-
tially confronts a world separated from that of its mother, on
whom it has depended to meet its every need, the child experiences a fragmented world. This experience of incompletion is
resolved as the child encounters its own reflection in the mirror.
Its mirror image (imago) restores a sense of coherency. The
imago synthesizes the internal chaos of libidinal energy a child
experiences in a unified external image. It may only be a mirror
image, but it performs an important function in the development of a child's ego: a "self" appears as the child actively

identifies with its distant reflection. The imaginary construct of a "self" is both the source of narcissistic misrecognition and a site of alienation.[27] For Lacan, a person's identity is basically an illusion formed through an act of misidentification.

An apocalyptic imagination justifies some of the most brutal instances of capitalist engagement and social conservatism since that imagination poses a singular problem, and herein lies its mythic power. Like the child who misidentifies with a unified image of itself in the mirror, an imago that reconciles different libidinal struggles into one coherent image, an apocalyptic imagination presents the encounter with the problem of crisis as having one solution. Crucially, the solution is not motivated by hopes for a different future. The solution doesn't promise any concrete changes; indeed, the myth of Armageddon gives up on futurity. And it is here where the narcissistic structure of the imaginary comes into play. Instead of confronting the existential crisis capitalism has created head-on, an apocalyptic imagination resolves the feelings of anxiety and powerlessness over the future that living under a violent socioeconomic system produces into an awe-inspiring image of natural disaster and redemption. Basically, the imaginary of apocalypse represents the feelings of powerlessness experienced under global capitalism in benign form. This obfuscates and helps prevent the "crisis" environmental degradation presents vis-à-vis global capitalism from being a real game changer.

For instance, under the threat of ecological collapse enviropreneurs (free market environmentalists) quickly move in under the umbrella of environmentalism turning ecological resources into environmental assets and environmental conservation into new free market opportunities, what Naomi Klein has fittingly described as "disaster capitalism."[28] Here a myriad of environmental problems are articulated on the singular horizon of advancing capitalism. The solution: mobilize capitalism to solve precisely the problem it has created. Moreover, it allows opportunists to cunningly deploy an exploitative and oppressive economic structure under an ethical guise. Or, replacing concerns for escalating climate threats with another threat: nuclear annihilation.[29] In this case the multifaceted problem of climate change is turned into another problem. Giving preference to one kind of annihilation over another is not a solution to annihilation.

It was Hannah Arendt who incisively remarked that mass society "wants not culture but entertainment, and the wares offered by the entertainment industry are indeed consumed by society just like any other consumer goods." "Entertainment, like labor and sleep," she added, "is irrevocably part of the biological life process."[30] Like Debord, Arendt regarded this kind of consumption as passive. Yet they were writing from a very different time from the hypertheater of immersive environments, special effects, and sonic manipulation entertaining the general public today. As Rancière has explained, hypertheater changes "representation into presence and passivity into activity."[31]

The apocalyptic imagination, as it is understood here, exhausts the political potential of imagining by tempting it with paralyzing awe and anguish. It operates at a prepersonal level amassing the energies of the social field in much the same way Wilhelm Reich described in *The Mass Psychology of Fascism*. Writing on European fascism in the early twentieth century, Reich explained that fascism was a social movement, populist in scope and appeal. To return to a point I explored in chapter 4, ideology is not one-sided, it isn't just a device political authorities use to hypnotize a stunned populace; the masses, in the case of National Socialism, actively wished for their own repression.[32]

Apocalyptic imagination makes its appearance as gratifying entertainment and an incurable cathartic sentiment that leads nowhere, to no one, and to no place and it is confined to the anxiety of a pure present, or what Žižek might call a subjective violence that is "experienced . . . against the background of a non-violent zero level" and "is seen as a perturbation of the 'normal,' peaceful state of things."[33] In essence, it marks that moment when the openness and free play of imagination are seduced by the hyperactive rehashing of short-tempered spectacles. Cornelius Castoriadis attributed this kind of imagination to "generalized conformism."[34] As Evans and Giroux have remarked, "extreme violence has become not only a commodified spectacle, but one of the few popular resources available through which people can bump into their pleasure quotient."[35]

The subject actively engages in forming an apocalyptic imagination as it closes off its interior life, like a lion circling its prey, slowly weakening all remnants of energetic fervor and public excitement to the point where the revolutionary potential of imagining is reduced to a jumbled mess of interiority. Basically, your own reality becomes the center of the world. Moreover, an apocalyptic imagination renders potentially political stimuli pleasurable, thereby sedating the revolutionary and transformative spirit of the political before it even comes into existence.

All that remains with an apocalyptic imagination are countless absolute presents that repeat the same way over and over again. This is why Chiara Bottici has claimed that spectacles have been drained of content, the effect of which is that "images are no longer what mediates our doing politics, but have become an end in themselves, which risk doing politics in our stead."[36] We have become "so image saturated" Bottici says, "it becomes increasingly difficult to create new images."[37] The vacuous production of spectacles Bottici identifies is a repetitive process that doesn't create the conditions from which critique can arise. Why? As Deleuze might have put it, this is an entirely closed activity of repetition without difference.[38] And what it repeats is not just a vacuous politics, a politics done through sound bites, quick edits, multiple angles, and "money shots" (in film, highly emotional scenes that grab a viewer's attention, and in photography a shot that drives sales). Repetition creates a competitive marketplace for political campaigns to vie for attention, in what Brown has described as "the undermining of democracy through the normative economization of political life."[39] Today, it is well known that airtime along with saturating social media outlets are crucial touchstones for the success of any political campaign.[40]

To help articulate the connection between politics and imagination, Bottici leans on the concept of imaginal beings: "We are not subjects who simply contemplate a world that is 'given,' but neither are we subjects who encapsulate the world within by consciousness. We are something in between the two. The problem is to understand what we are as 'imagining' beings."[41] As Bottici recognizes, the imaginal is not only oppressive but also carries an emancipatory potential, because the imaginal "can both open the path for critique, and thus autonomy, but also close it, as easily as it has opened."[42] In short, there is a direct connection between freedom and imagination.

On the freedom of the imagination, Immanuel Kant wrote: "In a judgment of taste the imagination must be considered in its freedom. This implies, first of all, that this power is here not taken as reproductive, where it is subject to the laws of association, but as productive and spontaneous (as the originator of chosen forms of possible intuitions)."[43]

Kant's insight in the first edition of *Critique of Pure Reason* is that imagination joins the distant happenings of intuition and thought. It plays a central role in the formation of knowledge by synthesizing the two. Knowledge, he maintained, arises when the raw data of intuition connects with the spontaneity of thought. Imagination synthesizes the apprehension of impressions with concepts to form a pure synthesis.[44] What he missed with his view of imagination as an individual faculty was the political potential of collective imagination, which emanates from the disruptive, not synthetic, operation of imagination. The irruptive and revolutionary potential of imagining negates synthesis.

For Castoriadis, the imaginary does not create images in either the psyche or the social field; it concerns signification. It is the radical potential of imagination where a subject's autonomy arises. A subject's autonomy depends upon the ability to break free from the constraints of actually existing circumstances and create a radically new (radical imagination) society ex nihilo. Imaginary presentations arise from an irreducible realm of prelinguistic, affective, and chaotic states. They are neither visual nor mere reflections of a preestablished sphere; rather, the imaginary is where meaning is produced from nothing.

Still, the lingering question is, how does imagining activate an emancipatory politics?

EMANCIPATORY IMAGINING

Emancipatory imagining breaks the linear connection that moves from *jouissance* to gratification. It does this by incorporating *jouissance* into human experience without fully quenching the thirst driving apocalyptic imagination. Why is this emancipatory?

Here imagination de-forms reality by realizing desire without completely satisfying it.[45] It is not a transcendent absolute event, of the kind that myth is made of; rather, the devastation and destruction environmental disasters inflict upon human beings, ecosystems, and other-than-human animals are part of every-day life.

Bottici has explained that "it is only by imagining the public that it exists."[46] The imaginal inflects and shapes how we apprehend the world. Unlike the concept of imagination that Kant developed, or the psychoanalytic concept of the imaginary, the imaginal refers to the collective and individual production of images, the content of which is not inevitably fictitious. In this regard, the imaginal is the product of the individual faculty of imagination and the social imaginary. It is the collective and individual understanding Bottici gives to the concept of the imaginal where she resonates with Arendt's understanding of imagination as political, one that harbors an emancipatory potential.

Arendt reminded us that the faculty of the imagination expands our worldview, allowing us to take on another's viewpoint; in this way imagination is central to realizing pluralism. Unsurprisingly, Arendt viewed the imagination as inherently political because it severs the ties of "identity-based experience," as Linda Zerilli has explained, through unrestrained free play.[47] Taking Arendt's connection between imagination and politics a step further, I don't regard imagining as inherently expansive. Indeed, imagining can be repressive; it all depends on how it is put to work, how the nonspecific intensive flows of energies and affects find investment.

Imagining engages sociality by creating new contexts of domination or emancipation; organizing and reorganizing how bodies, matter, and energy systems connect; and coding and decoding social structures. On the one hand there is a terribly repressive and fascistic operation of political imagining, one that can be tied to coding and a hierarchical logic of domination. On the other hand, the political promise of imagining is a relational movement that involves looking inward with critical reflection all the while maintaining an outward orientation that is open to difference. It does this by calling "forth forces in thought" and experience, which Deleuze might say "are not the forces of recognition . . . but the powers of a completely other model, from an unrecognised and unrecognisable *terra incognita*."[48] It is here where the political *promise* of an emancipatory operation of the imagination is stirred.

A promise reaches outward. It lunges the intimacy of the present beyond present circumstances. A promise presupposes change as much as it holds constant. The stability of a promise destabilizes the instability time presents. A promise kept is a promise recalled, extending back in time to a moment passed. Taking a walk in the dark, a promise lives inside trust. It is chosen by a golden heart of honesty and it rises to the fragrance of tomorrow. A promise remains incomplete, endlessly vulnerable; falling into the open arms of time, it swings along an arc of open responsibility. A promise assumes a gathering of forevers, darkening only when overshadowed by betrayal or stasis.

For Arendt, political space was constituted by a promise. It arises "directly out of the will to live together with others in the mode of acting and speaking."[49] A promise, in Arendt's view, is not founded upon the external authority or even the will of a single agent; it is a collective responsibility. A promise stirs forth a future held in common that is realized by remembering a past commitment.[50]

Writing on Arendt, Julia Kristeva has pointed out that a past functions to secure the future through what Nietzsche called a memory of the will, a distinction, which, for Arendt, "marks off human from animal life."[51] The promise functions as a "remedy for unpredictability, for the chaotic uncertainty of the future."[52] Arendt wrote, "Without being bound to the fulfillment of promises, we would never be able to keep our identities; we would be condemned to wander helplessly and without direction in the darkness of each man's lonely heart . . . a darkness which only the light shed over the public realm through the presence of others, who confirm the identity between the one who promises and the one who fulfills, can dispel."[53] According to Arendt, the "we" a promise institutes follows a relational logic. It depends upon the "presence and acting of others," and in this way it is an important constituent of political plurality.[54] However, Arendt's emphasis on a promise as instituting temporal continuity and future security strips the futurity out of the future. I am interested in another kind of political promise, one through which the unpredictability of the future provides the potential for alterity vis-à-vis the past and present.

Speaking with Arendt, I would agree that the power of a promise exceeds a single agent; it is inherently relational and the responsibility it invokes is infinite. I would add that over time a promise is repeated in different situations such that the original promise is reconstituted. In its repetition a promise necessarily exceeds the original promise such that the promise is independent of its maker(s). In this way, a promise calls forth a community yet to come. Conceived in this way, a promise is structured by the following three counterconditions: commonality, alterity, and openness.

I am reminded of a joke I once heard at the Passover table: Moses was feeling frustrated and fed up because Pharaoh and even the Israelites were angry with him. Then a loud, commanding voice from the heavens spoke to him:

"Moses, heed me! I have good news and bad news."

Taken by surprise, Moses perked up. The voice continued:

"Moses, you will lead the People of Israel from bondage. If Pharaoh refuses to release you, I will smite Egypt with a rain of frogs.

"Moses, you will lead the People of Israel to the Promised Land. And if Pharaoh tries to stop you, I will smite Egypt with a plague of Locusts.

"Moses, you will lead the People of Israel to freedom and safety. And if Pharaoh's army pursues you, I will part the waters of the Red Sea for you to cross to the Promised Land."

Amazed, Moses stammered, "That's . . . that's fantastic. I can't believe it! But what's the bad news?"

"Moses, you must write the Environmental Impact Statement."

The promise of a Promised Land brings with it a responsibility: use the land justly and peacefully. It is a promise that takes on unforeseeable responsibilities toward those who are living there—humans and nonhumans alike. God gives an infinite promise—the Promised Land—but the promise invokes an indefinite community, an unpredictable future that entails a responsibility to changing circumstances. A promise is inflected and revitalized through its repetition. The promise of a Promised Land is not exceptional; its significance arises in how it is used by redistributing our sensible "common habitat."[55]

Rancière has insisted that the political potential and limits of subjectivity are set through a distribution and redistribution of the sensible: "A 'common' world is never simply an *ethos*, a shared abode, that results from the sedimentation of a certain number of intertwined acts. It is always a polemical distribution of modes of being and 'occupations' in a space of possibilities."[56] As visibility and audibility are delineated through an organization of the sensible common world, the thinking and practice of political subjectivity are also established.

Now imagining becomes political. It reorganizes sociality through a redistribution of the sensible and a new time through which to "occupy the space of public discussions and take on the identity of a deliberative citizen" is opened up.[57] I regard the redistribution of the sensible that Rancière speaks of as a time of reimagining. It is also necessarily emancipatory precisely because it makes what was previously invisible and inaudible visible and audible. Put differently, emancipatory imagining presents a hypothetical situation that actualizes alternative modes of social organization.[58]

As Arendt amply demonstrated with her political concept of plurality as the underlying condition of intersubjectivity (fabrication, action, and speech), we collectively constitute a common world by sharing what we hold in common and our differences.[59] That is, intersubjectivity changes us. It expands our understanding of ourselves and enlarges our mentality. In effect, it pluralizes us. This is similar to the way in which Spinoza understood the imagination. Yet, unlike Arendt, the individual subject does not remain intact. Indeed, for Spinoza, the affective operation of the imagination works to "increase or aid the body's power of acting."[60] And this takes us into the terrain of posthumanist thinking.

As increasingly more and more species go extinct and the collapse of ecosystems becomes a transspecies challenge, the imaginative exchange between human subjects as the basis for politics is strained. Braidotti has taken this problem seriously, proposing that posthumanism offers a way out of the deadlock.[61] A posthumanist orientation creates a space and time through which the traumatic past of militarism, neoliberal governance, and global capitalism can appear in connection with contemporary ecological struggles and debates concerning the rights of future generations, other species, ecologies, and an expanded sense of "citizenship." By sharing what is held in common and through the differences that infuse and differentiate the common, shared realities are created, or what, in chapter 5, I call commoning. Commoning breaks apart fixed relations, compelling us to embrace flexible ontologies.

Yet flexibility is not inherently emancipatory; it all depends on how it works. Michael Hardt and Antonio Negri have aptly observed that confusing the "temporal division between work time and the time of life," the flexible labor economy of late capitalism is far more oppressive than it is liberatory.[62] Brad Evans and Julian Reid have also pointed out how under neoliberalism flexibility works as adaptation and resilience.[63] If we take our cue from Braidotti, the logic of flexibility central to posthumanism is rooted in a concept of affect; it changes us because it pluralizes us at an ontological level.

We are living in a new geological age, the age of the anthropocene. As human activities significantly impact and shape all life on earth, it has become increasingly urgent for people to critically reassess the exceptional position some humans have adopted vis-à-vis nature, other-than-human species, and other people. In light of this situation, Braidotti has mounted a compelling argument in favor of an affirmative politics of becoming posthuman. Using the philosophical apparatus offered by a combination of Spinozist and Deleuzian concepts such as affirmation, difference, and becoming, Braidotti pushes her critique of humanism beyond the contours of antihumanism and transhumanism. Posthumanism is not merely the biotechnological enhancement of humans. In this way, she insists upon the ethical charge that theoretical and practical efforts of becoming other than human offers. The challenge is simple—creatively reconsider human life with nonhuman others, such as nonhuman animal species, ecosystems, technology, and matter.

The spatial and temporal reciprocity of posthumanism is constituted by and in turn constitutes an event of commoning. It is not simply a modality that negates the human. Braidotti's work on posthumanism expands how imagination is employed. Posthumanism in the way she uses it becomes a method of analysis and an alternative strategy for social engagement. It generates new spatial and temporal connections. Above all, posthumanism destabilizes a subject's position in grammar and language as much as it smudges the contours of finite political identities. Identity politics rely, in large part, on a negative logic using distinct classifications such as species membership, class, race, gender, ethnicity, and sexuality to locate political subjectivities. A negative logic establishes what a person is based on what they are not. From here subjects are hierarchically distributed throughout the social field. Posthumanism disrupts the finite contours of identity and presumed primacy of the human over other species in all facets of life. Posthumanism, as I understand it, is a political orientation tensioned by a radical reimagining of the human figure as enriched, not compromised, by its connections with ecological systems, other-than-human species, and other people.

I returned to Paris for the COP21 talks (November to December 2015) to deliver a presentation as part of the UNESCO water and climate day. Although I understood the call for heightened security measures in the wake of the attacks, I was dismayed to learn the government had banned public demonstrations and protests. Undeterred by the prohibition, COP21 activists immediately changed tack under the leadership of the campaign group Avaaz by filling the Place de la République with their empty shoes. Among the shoes were those of UN secretary-general Ban Ki Moon and Pope Francis. Sneakers, loafers, sandals, flip-flops, high heels, and boots, some filled with plants, others with notes such as "S'il faut choisir un combat, c'est le climat" (If we are to choose a war, it is the climate).

A public square filled to the brink with vacant shoes formed a potent political image. A protest space was activated despite the absence of human bodies. This was a space of emancipatory imagining, a bodily action emptied of people as shoes defiantly occupied public space. The presence of shoes formed a dialectical connection with the absence of bodies and the injunction not to publicly protest. The power of imagination came from dialectically engaging with presence and absence and the memory of bodies through their excess.

The action awakened imaginary lines of comradeship with climate activists taking to the streets around the globe (November 29, 2015). To put it differently, the protest shoes in Paris combined different actions occurring around the globe in a unified image of solidarity, combining defiant impulses with activist ideas such as revolt and disobedience.

At the Palais de Tokyo, French philosopher Paul Virilio, American artists and architects Diller Scofidio + Renfro, Laura Kurgan, Mark Hansen, in collaboration with Native Land geographers and scientists, created an immersive 360-degree video installation depicting data collected between 2009 and 2014. The data portrayed fluctuations of the "world's population in motion" as a result of political, economic, and environmental forces. The installation consisted of a series of maps organized around six scenarios; each described a "component of these migrations."[64] For instance, population growth and decline, urban growth, global remittances along with distribution of remittances in relation to foreign aid, numbers of refugees and internally displaced persons, the growth in environmental refugees, sea-level rise, global changes in temperature, and global patterns of tree-cover loss caused by burning and logging. Through the use of animation and evocative soundscapes, *EXIT* drew out the affective dimensions of data, deepening the abstract through a procedure of numerical storytelling. The end result: through an exercise in emancipatory imagining, the particular was extracted from the general and the impersonal realms of calculation and computation were individuated.

The *Artists 4 Climate* exhibition also proceeded as planned. A tightly curated exhibition featuring internationally renowned artists was installed throughout Paris, occupying public places and mobilizing the popular imagination for action on climate change now. The exhibition raised funds by way of auctioning off thirty donated works in support of climate change adaptation projects and to combat desertification.

Here are a few snapshots: Disquieting images of industrial landscapes by photographer Edward Burtynsky presented the alarming scale and quasi-surreal impact human activities are having on the world. Pasted on billboards throughout the city, photographs by Gideon Mendel confronted pedestrians with images of a drowning world, exhibiting an unnerving combination of human vulnerability and strength. At the Place du Panthéon people tenderly and playfully embraced twelve blocks of ice. Olafur Eliasson had harvested eighty tons of ice from the Greenland glacier in Nuuk, which he then had transported by container ship and truck to Paris before unloading it into the public square. The ice blocks were left to slowly melt in full public view over a period of nine days. The Grand Galerie de l'Évolution featured a meticulous arrangement of marine specimens in glass cases. Creating a hospital for the Great Barrier Reef, as artist Janet Laurence described it, *Deep Breathing: Resuscitation for the*

Reef, cast an emotive spotlight onto the slow and methodical death of the world's largest coral reef system.[65] The rapid demise of the reef is both natural and political. As the waters warm, coral becomes stressed, prompting them to eject the food-producing algae they depend upon from their tissue, which in turn leads to coral bleaching. Equally important, the die-off is also the result of geopolitical and domestic power relations that stymie efforts to curb carbon dioxide emissions and climate change.

Taken together, the cultural interventions around COP21 drew attention to the manner in which a person is a mixture of unacknowledged histories, memories, materialities, energies, and drives. The human body is made up of numerous elements and practices, many of which are invisible and which we share with other species, plant life, and even the earth's crust. Each and every one of us is an ecological assemblage that is socially, culturally, economically, and politically mediated. Together these interventions confirmed Braidotti's avowal of a posthuman political figure that is both a unit of political analysis and site of political reimagining. Furthermore, the artworks disqualified official attempts to bring dissent to a standstill. These actions recouped the political edge of COP21 by turning the space of the city into a site of emancipatory imagining.

Emancipatory imagining renders the political vivid by activating incongruity into public life. It irritates the routine of consumption and the veneer of decency under which bureaucratic reforms are carried out. It rattles the cage of institutionalized practices and social norms by embracing the challenge posthumanism presents.[66] Emancipatory imagining rouses collective excitement geared toward change, pushing excited energies beyond the interior life of an individual or nation, such that fervor and excitement exist in relation to external realities. Emancipatory imagining forces the singular sensations of a life into conversation with the collective shadows of history and invokes a promise that extends into multiple futures. It thus marks a time of inexhaustible agitation surrendering the present to forensic examination by bringing it into conflict with itself.

Alas, the Paris Agreement signed by 195 world leaders fell seriously short of what is needed to keep global warming within the stated range of 1.5°C to 2°C. Despite the urgency of the situation— rising sea levels, melting ice caps, storms increasing in frequency and intensity, species extinction—the agreement is not scheduled to come into effect until 2020, and the pledges made to cut emissions are nonbinding. The agreement, for example, cannot stop U.S. president Trump from implementing his irresponsible and dangerous antienvironmental agendas. Trump entered office on January 20, 2017, with the stated commitment to withdraw from the Paris Agreement, roll back on environmental regulations, and recommence construction of the Dakota Access and Keystone XL pipelines. The Paris Agreement is powerless to prevent him from doing this, but COP21 countries could impose climate-related trade tariffs on the United States now that Trump has pulled the country out of the Paris climate agreement. COP21 was a successful diplomatic feat, with 195 countries signing on to the agreement. Now that the United States, alongside Syria and Nicaragua, is no longer part of the Paris Agreement, the greatest challenge moving forward is how to translate the emancipatory political imagining arising from the streets of Paris into political influence. Despite the environmental crisis pushing new solidarities to form among previously isolated political interests, the U.S. withdrawal from the Paris Agreement is a real blow to the climate alliance. The country is the second largest emitter of CO_2 in the world and historically is the largest carbon polluter in the world. The challenge that remains is, what political form is up to the gargantuan task of realizing the change emancipatory imagining presents?

9

SO TO SPEAK

As the earth's life systems face ecological collapse, the global inequity gap grows, the political cards stack in favor of the world's elite, and global capitalism and militarism create skewed power relations, the question of how social change will occur becomes increasingly urgent. Will the universal issue of environmental degradation produce a transformative political force that motivates people to traverse the identity boundaries of race, class, gender, sexuality, geography, age, physical ability, and speciesism? Will the transnational information age provide a strong enough platform on which this force can sufficiently mobilize an emancipatory movement— Twittering, e-mailing, Facebooking, and digging down deep into the web, discovering allies both near and afar? Will this new professionalized social movement infiltrate and occupy institutional spaces where government and policy making occur? Or will it be more anarchical, like the mass exodus Michael Hardt and Antonio Negri have described, an exodus that doesn't involve moving places but one that transforms the "relations of production and mode of social organization under which we live . . . an exodus from the relationship with capital and capitalist relations of production"?[1] Maybe it is as simple as refusing to identify with the current situation, as Slavoj Žižek has proposed. Or, will it be a more concentrated and situated effort targeting where capital is both produced and realized, taking place on the factory floor and the streets of the city, as David Harvey has maintained?[2] Or, do we need to reconceive the role of the party in political life, as Jodi Dean has so vividly and convincingly argued?[3]

Then there are those who are so marginalized that making the time to participate in political organizing means an impossible choice between putting food on the table or not. For a single mother working several shifts a day, taking the time to participate in the formation of a social movement may be a luxury she cannot afford. Disenfranchised groups are often not even at the table when it comes to deciding the direction of social, environmental, and economic policies. Remember the African contingency's walking out of the Copenhagen climate talks? They resorted to it because they realized their concerns were not being heard. Even among the ranks of social movements, many minorities remain silent and invisible. When thousands of Native Americans from across the United States marched on the White House March 10, 2017, to protest against President Trump's executive order reopening construction of the Dakota Access and Keystone XL pipelines, their efforts received little attention from both mainstream media and the president himself. It is disheartening to know that from 2000 to 2009 the majority of the US$10 billion for environmental activism funding in the United States went to professional organizations despite the fact that they failed to significantly influence environmental policy on Capitol Hill, not even cap-and-trade legislation, which is one of the most market-friendly of environmental policies.[4]

So we return to the problem of social change. How does an emancipatory politics arise when even some environmental groups join in on the neoliberal party? In Hardt and Negri's view, the world has qualitatively changed. State imperialism no longer holds sway. Instead, transnational institutional and economic powers are the major players. The center of power (state) has become a ubiquitous movement of networked power relations that infiltrate all aspects of social life.[5] They do, however, urge that a revolutionary future is inherent to the system of capital accumulation. Capital doesn't just produce exploitation and oppression but also creates a new common. Their use of the term *common* refers not only to natural resources but also to "those results of social production that are necessary for social interaction and further production, such as knowledges, languages, codes, information, affects."[6] And their revolutionary subject is the "multitude."

Their concept of the multitude moves beyond the unitary view of *the people* and the indifference of *the masses*, where all differences are submerged and drowned. The multitude is unconnected, yet it operates like a network. It is a multiplicity made up of a potpourri of singularities, expanding the Marxist political subject of class to include the "potentially infinite number of classes that comprise contemporary society based not only on class (however variegated that category is) but also on those of race, ethnicity, geography, gender, sexuality."[7] Using the South African protest slogan "We Are the Poors" as a foray into their revision of the unified political entity "we the people," the multitude galvanizes around the optimistic vision that "another world is possible."[8] Operating subversively, the multitude uses the globalized system of oppression against the grain, by, for example, putting the information economy to work to promote political change. Forming networked political activities and constituencies, the multitude flees its "relationship with the sovereign" and applies itself to creating a new world.[9]

The revolutionary politics Hardt and Negri offer might liberate the future from state power, yet the nagging question of to what degree their decentralized formula of the multitude changes the current course we are on persists. Isn't one of the key principles of neoliberalism to reduce government regulation? If global capitalism has seen the undermining of state power, as they attest, then wouldn't terminating the state reinforce the very nebulous system of power they so openly condemn? Are Hardt and Negri suggesting that the corporate sector would regulate itself? Ultimately, this is what landed us in the present mess.

Too often the question of a public and its revitalization comes down to cost: who will step up and subsidize the infrastructure and resources needed to sustain the public realm, such as public parks, streets, transportation, infrastructure, education, and health care? Unfortunately, all too often the public is forced to turn to the private sector for assistance because the state is so bankrupt. Accordingly, the public is placed in the service of private enterprise. So much for voluntary choices! The lesson to draw is that the choice between supposedly polar opposites—democracy and terror—is false. They are examples of what Žižek has called the ruling ideology imposing a "forced choice on us: we are free to choose only if we make the right choice ("democracy or terror"—who would choose terror?)."[10]

Hardt and Negri's suggestion to reject the state in favor of the open multitude operating transnationally might change the form of neoliberalism but possibly not its effects. Indeed, the revolutionary future of a transnational self-regulating multitude runs the risk of reinforcing the stranglehold corporate power currently has over public life. Simon Critchley has called this the "eternal temptation of anarchism."[11] In this sense their version of anarchy runs the risk of providing the perfect ideological supplement to neoliberal democracy and capitalism.

Meanwhile, Harvey has refused to give up on the central importance of political organization and the role of the state in public life. Faithfully Marxist in his dialectical reasoning, Harvey highlights the growing contradiction between social relations of production (social class producing and reproducing capital) and the forces of production; for him, the political project is clearly situated in the Marxist concept of class struggle, which he expands to include both working and urban classes. Workers might fight for higher wages and better working conditions, but urban dwellers are struggling to make ends meet paying rent, interest, and higher living costs. Harvey rightly points out that with the debt economy new sites of political struggle have arisen at the places of production and points of realization. More than half the world's population now live in urban centers, where the majority of value is also realized. Although labor might be outsourced to a rural factory in India, products are sold primarily in the urban sphere. For these reasons, Harvey maintains that the categories of class and urban dweller provide a newfound impetus for political life. This conclusion sits uncomfortably with 2016 election results. In the United Kingdom the success of the Brexit campaign lay with the disenfranchised rural voter.[12] In the United States rural counties ended up being the key ingredient in the popularization and subsequent election of right-wing presidential candidate Donald Trump.[13]

Despite Harvey's brilliant critique of capitalism, for him, capitalism harbors the potential for revolutionary change. That is, the Marxist position that capitalism produces its own contradictions and appropriates and places them in the service of further capital accumulation is also the same process that exacerbates and eventually realizes the revolutionary potential of class struggle.

How do Hardt and Negri's open flow of energies, affects, and forces eventually find investment in a revolutionary transnational political subject of the multitude? History has taught us that the investment of social energies, affects, and forces are not always emancipatory. What if Harvey's working classes and urban masses want to replace neoliberal capitalism with a more repressive system? What do we do then? We need only recall the failed experience of communism under Stalin, the fascist movement in Europe, and the Rwandan genocide to clearly make the point. When we add the desiring masses into the political mix, the revolutionary and progressive promise of the public is complicated. The ticklish question is, what if the masses desire their own oppression?

Cracks in Harvey's unified revolutionary underdog class appear with the realization that not everyone in the working and urban classes is necessarily interested in advancing emancipatory and progressive political agendas, let alone radical ones. In the United States the reactionary politics of the New Right, like the Tea Party or the U.S. President Trump and his Republican administration, have garnered a great deal of support from the working poor, more specifically conservative white men. The problem lies with the presupposition that revolution entails a taking of power without interrogating the configuration of the movement itself.

The revolutionary activities of the multitude are less a matter of taking power from the ruling class than the undermining of the triadic system of authoritarian power—nation, family, corporation. In their work, Hardt and Negri tackle the challenge social desire presents for a revolutionary political agenda by proposing the multitude resist authoritarian power on all three fronts (nation, family, corporation). Hardt and Negri's suggestion is for the multitude to skirt sovereign power by giving it the "capacity of democratic decision-making."[14] But how do we get from the authoritarian system of sovereign rule to the openness of the multitude? Žižek has aptly pointed out that, despite all their good intentions, this is quite the "qualitative jump" from the "multitudes resisting the One of sovereign Power to the multitudes directly ruling themselves."[15]

In contrast, Dean has offered an exciting way forward for left-leaning politics. She argues for the importance of forming solidarities, all the while recognizing that the "we" at the basis of this is perpetually changing. Solidarities work through antagonism, challenging the status quo and the formalized political domain. Skirting the trap of identity politics, she reimagines the political potential of the crowd and the use value of the party as a "political association that cuts across workplace, sector, region, and nation."[16] Her view of revitalizing the communist organization of the party form is not a return to the authoritarian party structures of Maoist China or Stalinist Russia. Rather, she hits the refresh button on the Communist Party, infusing a principle of "collective voluntary cooperation" into it.[17]

Considered this way, the Communist Party is "an organization premised on solidarity, the party holds open a political space for the production of a common political will, a will irreducible to the capitalist conditions in which the majority of people find themselves forced to sell their labor power. Where work is obligatory, membership in the party (like participation in the movement) is voluntary, the willed formation of united power. Among its members, the party *replaces* competition with solidarity."[18] The solidarity of the people, however, does not point to a unified whole, nor the general consensus common to the democratic view of the people as represented in a parliamentary or bicameral system. Communism presumes the people are divided and the party, says Dean, is a form that asserts this split. There is always an excess to democracy hovering in the background of neoliberal self-serving claims of self-interested freedom, choice, and individualism. Jacques Rancière has called this the invisible, inaudible minority. Dean describes it as the "intensity of the egalitarian discharge."[19]

When the demands of the crowd, such as the protesters at the UN climate talks, fall on deaf ears, a blind spot in the representative frame of democracy is exposed. What we are dealing with here is the self-contradictory nature of neoliberal democracy. A radical negativity at the core of the democratic project arises when the two political signifiers—reformist or revolutionary—cannot be simultaneously present, so the secondary signifier (the revolutionary) is repressed. Sometimes the repressed signifier erupts and returns to haunt the first. That haunting can come in many different sizes, shapes, and forms. It can be fascistic or emancipatory, reactionary or egalitarian, intolerant or progressive—it all depends on how the energies and affects of the social field are invested and the series of signifiers that return to fill in the gap between the two master signifiers.[20]

In *Anti-Oedipus*, Gilles Deleuze and Félix Guattari highlighted the importance of addressing the libidinal charge of social life and the problem desire presents for progressive politics. Desire, for them, is productive and social. It is a connective process that brings together energies, affects, and forces, combining to produce a social system that is neither "expressive or representative, but productive."[21] This combination of energy, affects, and forces either finds investment in despotic, fascistic, authoritarian social arrangements or it deterritorializes fixed relations, releasing differentiating lines of flight that create inclusive and open social arrangements. The same holds true for Hardt and Negri's conceptualization of the multitude, who, they say, operate along lines of flight, what Deleuze and Guattari draw attention to in their analysis of desire as social. This is the "so to speak" of the public's role in political life—the public can go either way, in a reactionary or a liberatory direction; it all depends on how social energies and forces are configured.

While activist politics are continually being undermined by corporate government, big business, and militarized social life, presenting a serious challenge for transformative political agendas, so, too, do the shared anxieties environmental crises incite. How can a common experience of hardship and suffering occupy center stage in environmental activism without the anxiety that hardship creates driving the character structure of the movement? Anxiety presents a chaotic and indeterminate charge into the social field, operating like the real does, in Žižek and Jacques Lacan's thinking. The anxiety produced by crises resists symbolization; it thereby maintains an ontological status of a priori excess. However repressed they might be, the anxieties climate change and ecological collapse present to society return as a symptom that is both enjoyed and hated at the same time. Why else does the mainstream media barrage the public with images of natural disasters. The spectacle of calamity strikes a chord that runs deep throughout the social field, terrifying as much as it is pleasurable.

Climate change and ecological collapse force the social field to nudge up close to the "unfathomable X which forever eludes the symbolic grasp," to borrow from Žižek.[22] The reference is to, namely, the manner in which both neoliberal economy and corporate government are the collective expression of authoritarian drives operating throughout the underbelly of the social field. How else does disaster capitalism work, as Naomi Klein has so aptly called it, if not by replacing the unknown angst crisis creates with authoritarian initiatives such as economic restructuring and austerity and by disciplining government spending. Put differently, economic rationalizing, corporate governance, and militarism institutionalize the excessive and unmanageable facets of political life, and along with it the public gains a false sense of security. This is the central point of my criticism of favela development projects in *Hijacking Sustainability* (2009): as the streets were paved and broadened, not only did they allow the trash to be collected but they also facilitated the militarization of favela spaces as the state police and military could swiftly enter the area, and along with that came the criminalization of the poor.

Considered this way, a different kind of gap from Harvey's working urban class versus capitalist-corporate government or Hardt and Negri's transnational multitude transforming the authoritarian system of capital from within the system emerges. It involves pushing the materialism at the core of their work to the point where political awareness arises from the realization that "the reality I see is never 'whole'—not because a large part of it eludes me, but because it contains a stain, a blind spot, which indicates my inclusion in it."[23] We now encounter a gap at the very heart of contemporary political experience that arises from the psychosocial configuration of the public. Erich Fromm described this channeling of individual human energies as a productive social force, clarifying in *Escape from Freedom* a common misunderstanding that Hitler took over the state through "trickery and cunning" when in fact "millions in Germany were as eager to surrender their freedom as their fathers were to fight for it; that instead of wanting freedom, they sought for ways of escaping from it."[24]

In the final pages of *Welcome to the Desert of the Real* Žižek proclaims democracy is currently in a deadlock. The democratic event both eschews materialist causality and functions as the basis of historicity (the emergence of the new). Instead, the populist Right, he explains, "acts, sets the pace, determines the problematic of the political struggle, and the liberal centre is reduced to a 'reactive force'" plundering the public along with it.[25] Philosophically speaking and borrowing from Žižek, one step in this direction begins by discerning the non-all of a public—that is, the "void that separates" the "material reality" of a public "from itself."[26] We cannot fully grasp the political function of the public without engaging the "Real of the speculative dance of Capital."[27]

The lingering question is, how can a public engage the real that lies outside neoliberal language without reverting to an immature state of dependency and start growing up and embracing freedom for all its anxieties, joyfulness, and autonomy? This involves constructing a "new reality: the adulthood of the species," as Thomas Fisher has implored us to do.[28] Or, as Žižek contends, writing on the ecological crisis as the "answer of the real" in *Looking Awry*, "The only proper attitude is that which fully assumes this gap as something that defines our very *condition humaine*, without endeavoring to suspend it through fetishistic disavowal, to keep it concealed through obsessive activity, or to reduce the gap between the real and the symbolic by projecting a (symbolic) message into the real. The fact that man is a speaking being means precisely that he is, so to speak, constitutively 'derailed,' marked by an irreducible fissure that the symbolic edifice attempts in vain to repair."[29] Inevitably the political promise of environmental politics rests with emancipatory imagining, to imagine new and unforeseeable realities, so as to exercise the anxieties this might generate and redirect them away from authoritarian and oppressive social systems and toward the construction of emancipatory and egalitarian futures.

The political here indicates a "radical risk" and the "madness of a decision" that Žižek describes in a nod to Derrida and Kierkegaard.[30] An act that breaks free of the limits of capitalist democracies, which foreclose the political act through "democratically legitimized" social change.[31] Instead, Žižek advocates for "terrifying excess," breaking free from the empty posturing of liberal democratic politics. Climate change and ecological collapse are that terrifying excess Žižek speaks of.

Gangster corporatism, governance, and militarism are plunging the transformative potential of environmentalism into crisis. Environmental politics needs to be revitalized and infused with a newfound sense of activist hope, but a leap of faith is not going to get us there. Arundhati Roy's plea in *The Cost of Living* is instructive here, for "as long as we have faith, we have no hope. To hope, we have to *break* the faith." Pressing on, she advises, "We have to fight specific wars in specific ways. We have to fight to win."[32] And she is absolutely right. Because if we lose the battle the private sector, corporate governance, and militarism are waging against life on earth, we lose the right to breathe fresh air, quench our thirst with clean water, and even to look our children in the eyes and state with absolute conviction that we have done everything in our power to leave behind us a world they and other generations will find worth living in.

The first battle begins at the level of the imagination and ideas. Personal politics, the citizen consumer, and militarism have wiped out civic responsibility and cooperation from the language of mainstream politics. A diabolical faith in individualism has fractured us as a society. The sensuality of human beings has been reduced to the narcissistic pleasures of spectacle culture and entertainment. An aesthetic of simulacra allays the fears autonomy incites, allowing the human race to narcissistically immerse itself in versions of itself, where nothing or no one is ever truly separate from this egocentric fantasy. As such sensitivity toward another's suffering or care over another's well-being is never experienced on its own terms, it is mediated by a self-aggrandizing psychosocial ego. The very idea of showing responsibility to future generations, other species, or even less-fortunate members of our own species is thus foreclosed before it even germinates.

Political projects need a common ground. The persistent problem is how the central status of the common in political life appears. How can the energies and affects of the social field invest in inclusive, open, and creative political trajectories ahead of the authoritarian impulse of reactionary politics? This ultimately comes down to imagining. Without emancipatory imagining, all that remains is the parched terrain of intractable dogma and categorical belief. Imagining arouses feelings of uncertainty, awakens surprise amid the everyday, leaves us helpless, enflames the creative temporality of remembrance and forgetting, stirs the affective potential of perception, and provokes unfamiliar sensations using familiar emotions. Imagining speaks, so to speak, giving voice to aspiration, anticipation, audacity, fragility, and chance. Never fully fastened and sealed, emancipatory imagining is a movement that releases the senses, breaking down social roles, scratching away at supposedly trivial details, suspending meaning, acquiring unfamiliar feelings, proceeding by seduction and nausea, all the while arousing perception and passion. All find their climax in a bric-a-brac of unintended fresh ideas and actions that "come about 'by themselves.'"[33]

We need to revive the collective political subject—to build a solidarity movement that arises from emancipatory imagining, to pursue egalitarian political agendas, and institute a strong political position, one that begins by drawing a clear line in the sand: capitalism, neoliberal governance, and militarism are incompatible with social flourishing.

At the time of this writing the world was undergoing a massive change in political climate. New, lively forms of left-wing politics had arisen in Spain with Podemos and in Greece under Syriza. Meanwhile, the influx of refugees across the waters of the Mediterranean were threatening to rip apart the political fabric tenuously tying the European Union together. The Paris attacks marked a victory for the right wing. And it is not only France where the right wing is making a comeback. The growing support for anti-immigration nationalist politics across the European Union is alarming. It includes Fidesz and Jobbik in Hungary; the Polish Law and Justice Party, which won the presidency and a parliamentary majority; GERB and Ataka in Bulgaria; and the Freedom Party of Austria (which came very close to winning the presidency, losing by only a 0.6 percent margin). The Brexit campaign won the referendum in the United Kingdom. As a result, Scotland was threatening to leave the United Kingdom because it wanted to remain in the European Union. In the United States a misogynistic and racist presidential campaign took the country by storm, garnering the worst of public sentiment and intolerance under the figurehead of Donald Trump. And all across the world far-right-wing religious groups and individuals have shot, bombed, raped, attacked with machetes, and taken ordinary people hostage regardless of age, gender, nationality, class affiliation, or sexual orientation.

Meanwhile, another violence moves quickly and quietly amid the war and divisive politics of everyday life, and it is best summed up by the following statistic: as of January 2016 sixty-two billionaires owned half the world's wealth. Moreover, despite the global population's increasing by approximately 400 million people since 2010, the wealth of the poorest 3.6 billion people has dropped 38 percent, estimated at $1 trillion.[34]

Amid a world immersed in violence, everyday people continue to cope with massive flooding, heat waves, extreme storms, poisoned water supplies, insufficient sanitation, indebtedness, not to forget life in a perpetual war zone. And it is the poor, women, ethnic and sexual minorities, and other-than-human species that are the most vulnerable. Environmental emergencies make these social fault lines all the more acute.

The political arena is filled with complexity and ambiguity, and a great deal of the shift to the right can be attributed to the failure of visionary leadership. Politics contributes to social change, and at its best it helps societies rise to the challenge that new social and environmental pressures and struggles present. I have argued in this book that, despite the weakening of the emancipatory promise of environmentalism, environmental politics is not just an issue-based social movement. It is emancipatory and egalitarian, bringing together different constituencies to situate political life in the gaps between these, improvising with distributive centers of power.

Indeed, environmentalism arises from and complements a variety of social justice movements but is not bound by an anthropocentric understanding of the social. It interrogates how politics is practiced by working in partnership with animal rights activists, conservationists, human rights advocates, and refugee action collectives, using a posthumanist orientation, in the sense that Rosi Braidotti has described posthumanism.

The politics of environmentalism is agility. The various movements work affirmatively within the dominant system with a view to changing it, only to quickly shift gears to avoid co-optation, adopting a negative strategy of confrontation and opposition. It appeals to the singularity of local experiences, resources, capacities, and histories all the while embracing a universal attitude that aspires for the dignity and flourishing of all life on earth both today and into the future.

This is the emancipatory and egalitarian promise environmentalism presents life in the twenty-first century.

AFTERWORD

We witness
bitter poverty and persecution
driving people into the waves
their thirsty bodies flood
another's shoreline.
A shoreline to be shared
but which no one dares.
A new life intercepted
and turned away.
And their darkness is panic.

We witness
waters monstrous and thick,
ironclad bodies
navigate the riptides
filled with viciously good intentions
not upset
without regret
all hoping to protect
a national story

from foreign spaces of emptiness.
And their darkness is ferocious.

Imagine
a people neither blindfolded
nor kidnapped
by fundamentalists,
opportunists,
terrorists,
or
capitalists.
Standing in defiance amidst the dust and debris
of hate speech
and gangsters,
outlawed by compromises
made in other lands.
And the darkness falls silent.

Imagine
the stature
of a species never hunted
looking its killer straight in the eyes.
Or
a warrior dragging
the secrets of war
through the trapdoor.
And the darkness, it's booming.

Imagine
another world
of wind, waves, and sun
energizing a planet once in waiting,

announcing: *tomorrow has come.*
Another world
impenetrably open,
struck by tomorrow's reach
it throws out a mischievous wink,
linking arms in deep
friendship.
Like a young girl's dreams
released inside a woman's body,
tomorrow finds its way
past overcast skies
and mobster lies.
Filled with laughter,
the kind that saturates a world
with the sweet sounds of sisterhood and fireflies,
it swaps the forsaken futures of yesterday
where hustlers roamed the beaten tracks alone
and the hungry stared silently through the heavy heat,
with bright-eyed conviction and wonder.

Today, strong steady lips embrace
the promise tomorrow brings.
Today, together, and unrelentingly tender
the earth stands unashamed and tall
stealing a long overdue kiss
from which new earth is born.

NOTES

INTRODUCTION

1. Steve Cole and Leslie McCarthy, "Long-Term Warming Likely to Be Significant Despite Recent Slowdown," *NASA*, March 11, 2014, accessed March 22, 2014, http://climate.nasa.gov/news/1050.

2. Jonathon Pearlman, "100,000 Bats Fall Dead from the Sky During a Heatwave in Australia," *Daily Telegraph*, January 8, 2014, accessed March 1, 2014, http://www.telegraph.co.uk/news/worldnews/austra liaandthepacific/australia/10558183/100000-bats-fall-dead-from-the -sky-during-a-heatwave-in-Australia.html.

3. Jean-Marie Robine, Siu Lan K. Cheung, Sophie Le Roy, Herman Van Oyen, Clare Griffiths, Jean-Pierre Michel, and François Rich- ard Herrmann, "Death Toll Exceeded 70,000 in Europe During the Summer of 2003," *Comptes rendus biologies* 331, no. 2 (2008): 171–78.

4. Thomas Tidwell, "Wildland Fire Management," *Statement before the Committee on Energy and Natural Resources, U.S. Senate*, June 4, 2013, accessed March 1, 2014, http://www.energy.senate.gov/public/index .cfm/files/serve?File_id=e59df65c-09c6-4ffd-9a83-f61f2822a075.

5. Debbie Hillier and Benedict Dempsey, *A Dangerous Delay: The Cost of Late Response to Early Warnings in the 2011 Drought in the Horn of Africa*, Oxfam International and Save the Children, January 18, 2012, accessed March 28, 2017. http://policy-practice.oxfam.org.uk/publica- tions/a-dangerous-delay-the-cost-of-late-response-to-early-warnings -in-the-2011-droug-203389.

6. Slavoj Žižek, *Welcome to the Desert of the Real* (London: Verso, 2002), 149.

7. UNHCR, *Global Trends: Forced Displacement in 2015*, accessed August 9, 2016, https://s3.amazonaws.com/unhcrsharedmedia/2016/2016 -06-20-global-trends/2016-06-14-Global-Trends-2015.pdf.

8. Tim Herzog, "World Greenhouse Gas Emissions in 2005," World Resources Institute, July 2009, accessed March 1, 2014, http://www .wri.org/publication/world-greenhouse-gas-emissions-2005.

9. Slavoj Žižek, *The Parallax View* (Cambridge, Mass.: MIT Press, 2006), 384.

10. Slavoj Žižek, *Organs without Bodies* (Oxon: Routledge, 2004), 24.

11. The etymological root of *democracy* is Greek *demos*, meaning "people," and *kratia*, meaning "power." When combined, it refers to "rule by the people."

12. Susan Sontag makes a similar argument in her analysis of science fiction movies ("The Imagination of Disaster," *Commentary*, October 1, 1965, 42–48).

1. VARYING SHADES OF GREEN

1. World Bank, *Global Economic Prospects 2007: Managing the Next Wave of Globalization* (Washington, D.C.: International Bank for Reconstruction and Development / World Bank, 2007), 30.

2. United Nations, *World Economic and Social Survey 2013: Sustainable Development Challenges*, accessed February 6, 2014, http://sustain abledevelopment.un.org/content/documents/2843WESS2013.pdf.

3. John Bellamy Foster, Brett Clark, and Richard York, *The Ecological Rift: Capitalism's War on the Earth* (New York: Monthly Review Press, 2010), 45.

4. James O'Connor, *Natural Causes: Essays in Ecological Marxism* (New York: Guilford Press, 1998).

5. Karl Marx, *Capital: Volume 1*, trans. Ben Fowkes (London: Penguin, 1990), 247–57.

6. Ibid., 248–49.

7. Annie Leonard, *The Story of Stuff: How Our Obsession with Stuff Is Trashing the Planet, Our Communities, and Our Health* (New York: Free Press, 2010); Raj Patel, *Stuffed and Starved: The Hidden Battle for the World Food System* (New York: Marble House, 2008).

8. Corporate Eco Forum, *The New Business Imperative: Valuing Natural Capital* (2012), 5, accessed January 4, 2014, http://www.corporateecoforum .com/valuingnaturalcapital/offline/download.pdf.

9. "Changes in Atmospheric Carbon Dioxide, Methane, and Nitrous Oxide," in *IPCC Fourth Assessment Report: Climate Change 2007*, accessed March 17, 2017, https://www.ipcc.ch/publications_and_data /ar4/wg1/en/tssts-2-1-1.html.

10. National Oceanic and Atmospheric Administration, *Up-to-Date Weekly Average CO_2 at Mauna Loa*, 2013, accessed June 1, 2014, http:// www.esrl.noaa.gov/gmd/ccgg/trends/weekly.html.

11. Leonard F. Konikow and Eloise Kendy, "Groundwater Depletion: A Global Problem," *Hydrogeology Journal* 13, no. 1 (March 2005): 317–20.

12. National Resources Defense Fund, "The BP Oil Disaster at One Year," *National Resources Defense Fund* (2011), accessed April 2, 2014, http://www.nrdc.org/energy/bpoildisasteroneyear.asp.

13. Eric Chivian and Aaron Bernstein, eds., *Sustaining Life: How Human Health Depends on Biodiversity* (New York: Center for Health and the Global Environment/Oxford University Press, 2008); David Tilman, Robert M. May, Clarence L. Lehman, and Martin A. Nowak, "Habitat Destruction and the Extinction Debt," *Nature* 371 (September 1994): 65–66.

14. Nik Heynan, James McCarthy, Scott Prudham, and Paul Robbins, eds., *Neoliberal Environments: False Promises and Unnatural Consequences* (London: Routledge, 2007), 11–13.

15. Andrew Blowers, "Transition or Transformation? Environmental Policy Under Thatcher," *Public Administration* 65, no. 3 (1987): 281.

16. Ibid., 282.

17. Ibid., 286.

18. Wilfrido Cruz and Robert Repetto, *The Environmental Effects of Stabilization and Structural Adjustment Programs: The Philippines Case* (Washington, D.C.: World Resources Institute, 1992); Susan George, *A Fate Worse Than Debt: The Financial Crisis and the Poor* (New York: Grove Press, 1988); Henry Owusu, "Current Convenience, Desperate Deforestation: Ghana's Adjustment Program and the Forestry Sector," *Professional Geographer* 50, no. 4 (November 1998): 418–36.

19. Owusu, "Current Convenience," 425–29.

20. Ibid., 426.

21. Center for Public Integrity and International Consortium of Investigative Journalists, *Promoting Privatization*, accessed May 5, 2014, http:projects.publicintegrity.org/water/report.aspx?aid=45.

22. As cited in Inga Schlichting, "Strategic Framing of Climate Change by Industry Actors: A Meta-Analysis," *Environmental Communication* 7, no. 4 (December 2013): 501.

23. Ibid.

24. As cited in Paul Reynolds, "Kyoto: Why Did the US Pull Out?" *BBC News*, March 30, 2001, accessed January 3, 2014, http://news.bbc.co.uk/2/hi/americas/1248757.stm.

25. Kirsty Hamilton, *The Oil Industry and Climate Change: A Greenpeace Briefing* (Greenpeace International, 1998), 35, accessed January 4, 2014, http://www.greenpeace.org/international/Global/international/planet-2/report/2006/3/the-oil-industry-and-climate-c.pdf.

26. John Vidal, "Revealed: How Oil Giant Influenced Bush," *Guardian*, June 8, 2005, accessed December 20, 2013, http://www.theguardian.com/news/2005/jun/08/usnews.climatechange.

27. David Suzuki, "Carbon Offsets as a Tool in the Fight Against Global Warming," *David Suzuki Foundation*, 2009, accessed April 10, 2014, http://www.straight.com/news/david-suzuki-carbon-offsets-tool-fight-against-global-warming.

28. As cited in Greenpeace, "U.S. Withdraws from Kyoto Protocol," April 5, 2001, accessed January 3, 2014, http://www.greenpeace.org/usa/en/news-and-blogs/news/u-s-withdraws-from-kyoto-prot/.

29. Foster, Clark, and York, *The Ecological Rift*, 71–72.

30. Ibid., 70.

31. Paul Hawken, Amory Lovins, and Hunter Lovins, *Natural Capitalism: Creating the Next Industrial Revolution* (Boston: Little, Brown, 1999), 319.

32. William McDonough and Michael Braungart, *Cradle to Cradle: Remaking the Way We Make Things* (New York: North Point Press, 2002).

33. Adrian Parr, *Hijacking Sustainability* (Cambridge, Mass.: MIT Press, 2009), 16–19; Adrian Parr, *The Wrath of Capital: Neoliberalism and Climate Change Politics* (New York: Columbia University Press, 2013), 130–44.

34. Parr, *The Wrath of Capital*, 8–38.

35. Ibid., 59.
36. Foster, Clark, and York, *The Ecological Rift*, 47.
37. Parr, *The Wrath of Capital*, 33.
38. Frankfurt School of Finance and Management and UNEP Collaborating Centre for Climate and Sustainable Energy Finance, *Global Trends in Renewable Energy Investment 2012*, 5, accessed January 1, 2014, http://fs-unep-centre.org/sites/default/files/publications/global trendsreport2012.pdf.
39. Thomas Picketty, *Capital in the Twenty-First Century* (Cambridge, Mass.: Belknap Press, 2014).
40. Matthew Miller and Peter Newcomb, "The World's 200 Richest People," *Bloomberg*, November 8, 2012, accessed January 8, 2014, http://www.bloomberg.com/news/2012-11-01/the-world-s-200-richest -people.html.
41. Oxfam, *Oxfam Media Briefing* (ref. 02/2013), 2, January 18, 2013, accessed January 8, 2014, http://www.oxfam.org/sites/www.oxfam.org /files/cost-of-inequality-oxfam-mb180113.pdf.
42. Naomi Klein, *The Shock Doctrine: The Rise of Disaster Capitalism* (New York: Picador, 2007), 530.
43. Nicholas Stern, *The Economics of Climate Change: The Stern Review* (Cambridge: Cambridge University Press, 2007), vi.
44. The professionalization of the environmental movement is part of a larger trend among social movements during the latter part of the twentieth century. See Larry Buffington, "Professionalization: A Strategy for Improving Environmental Management," *Humboldt Journal of Social Relations* 2, no. 1 (fall/winter 1974): 18–21; Craig Jenkins and Craig Eckert, "Channeling Black Insurgency: Elite Patronage and Professional Social Movement Organizations in the Development of the Black Movement," *American Sociological Review* 51, no. 6 (December 1986): 812–29; John McCarthy and Mayer Zald, *The Trend of Social Movements in America: Professionalization and Resource Mobilization* (Morristown, N.J.: General Learning Press, 1973); Suzanne Staggenborg, "Coalition Work in the Pro-Choice Movement: Organizational and Environmental Opportunities and Obstacles," *Social Problems* 33, no. 5 (June 1986): 374–90.

45. Parr, *Hijacking Sustainability*, 15–31; Foster, Clark, and York, *The Ecological Rift* 377–99.

46. Klein, *The Shock Doctrine*, 522.

47. Quoting Kumari, ibid., 492.

48. Judith Dean, "Does Trade Liberalization Harm the Environment? A New Test," *Canadian Journal of Economics* 35, no. 4 (2002): 819–42; Boqiang Lin and Chuanwang Sun, "Evaluating Carbon Dioxide Emissions in International Trade of China," *Energy Policy* 38, no. 1 (2010): 613–21; Christopher Weber, Glen Peters, Dabo Guan, and Klaus Hubacek. "The Contribution of Chinese Exports to Climate Change," *Energy Policy* 36, no. 9 (2008): 3572–77.

49. Jintai Lin, Da Pan, Steven J. Davis, Qiang Zhang, Kebin He, Can Wang, David G. Streets, Donald J. Wuebbles, and Dabo Guan. "China's International Trade and Air Pollution in the United States," *PNAS* 111, no. 5 (February 4, 2014): 1736–41, accessed January 4, 2015, http://www.pnas.org/content/early/2014/01/16/1312860111.full.pdf.

50. Parr, *The Wrath of Capital*, 14.

2. GREEN GOVERNMENTALITY

1. Michel Foucault, "About the Beginning of the Hermeneutics of the Self," *Political Theory* 21, no. 2 (1993): 204.

2. Lee Myung-bak, "A Great People with New Dreams," *Presidential Speeches*, August 15, 2008, accessed June 8, 2014, http://www.korea.net/Government/Briefing-Room/Presidential-Speeches/view?articleId=91000&pageIndex=9.

3. James Hansen, *Storms of My Grandchildren: The Truth About the Coming Climate Catastrophe and the Chance to Save Humanity* (New York: Bloomsbury, 2011); Nicholas Stern, *The Economics of Climate Change: The Stern Review* (Cambridge: Cambridge University Press, 2007); Michael Bloomberg, Henry Paulson, and Thomas Steyer, *Risky Business: The Economic Risks of Climate Change in the United States*, 2014, accessed June 28, 2014, http://riskybusiness.org/uploads/files/Risky Business_PrintedReport_FINAL_WEB_OPTIMIZED.pdf.

4. Lara P. Clark, Dylan B. Millet, and Julian D. Marshall, "National Patterns in Environmental Injustice and Inequality: Outdoor NO_2 Air Pollution in the United States," *PLOS ONE* 9, no. 4 (April 2014):

1–8; Tânia Pacheco, "Inequality, Environmental Injustice, and Racism in Brazil: Beyond the Question of Colour," *Development in Practice* 18, no. 6 (November 2008): 713–25.

5. Robert Bullard, ed., *Just Sustainabilities: Development in an Unequal World* (Cambridge, Mass.: MIT Press, 2003); Robert Bullard, *The Quest for Environmental Justice: Human Rights and the Politics of Pollution* (San Francisco: Sierra Club Books, 2005); United States Environmental Protection Agency, "Environmental Justice Program and Civil Rights," 2014, accessed October 10, 2014, http://www.epa.gov/region1/ej/.

6. Thomas Picketty, *Capital in the Twenty-First Century* (Cambridge, Mass.: Belknap Press, 2014).

7. Organisation for Economic Co-operation and Development, *Towards Green Growth* (2011), 13, accessed May 10, 2014, http://www.oecd.org/dataoecd/37/34/48224539.pdf; see also World Bank, *Inclusive Green Growth: The Pathway to Sustainable Development*, 2012, accessed May 10, 2014, http://issuu.com/world.bank.publications/docs/9780821395516.

8. Economic and Social Commission for Asia and the Pacific, Asian Development Bank, and United Nations Environment Program, *Green Growth, Resources and Resilience: Environmental Sustainability in Asia and the Pacific*, 2012, accessed June 6, 2014, https://www.adb.org/sites/default/files/publication/29567/green-growth-resources-resilience.pdf; World Bank, *Inclusive Green Growth*.

9. Paul Ekins, *Economic Growth, Human Welfare and Environmental Sustainability: The Prospects for Green Growth* (London: Routledge, 2000); Lee Myung-bak, "Great People"; Stern, *Economics of Climate Change*.

10. Economic and Social Commission for Asia and the Pacific et al., *Green Growth*, iii.

11. Ibid., xii; Martin Ravallion. "How Long Will It Take to Lift One Billion People Out of Poverty?" *World Bank Research Observer* 28, no. 2 (August 2013): 139–58.

12. World Bank, *Inclusive Green Growth*.

13. Economic and Social Commission for Asia and the Pacific et al., xvi.

14. Ibid., 16.

15. Wendy Brown, *Undoing the Demos: Neoliberalism's Stealth Revolution* (New York: Zone Books, 2015); James O'Connor, *Natural*

Causes: Essays in Ecological Marxism (New York: Guilford Press, 1998), 234–51; Adrian Parr, *The Wrath of Capital: Neoliberalism and Climate Change Politics* (New York: Columbia University Press, 2013), 19–21.

16. Picketty, *Capital*, 25.

17. Ibid.

18. Ibid., 23.

19. Ibid., 26.

20. Ibid., 25.

21. Ibid., 1.

22. David Harvey, *A Brief History of Neoliberalism* (Oxford: Oxford University Press, 2005), 187.

23. Naomi Klein, *The Shock Doctrine: The Rise of Disaster Capitalism* (New York: Picador, 2007).

24. Jamie Peck, *Constructions of Neoliberal Reason* (Oxford: Oxford University Press, 2010), 28; Philip Mirowski, *Never Let a Serious Crisis Go to Waste* (London: Verso, 2014), xi.

25. Harvey, *Brief History*, 183.

26. Mirowski, *Serious Crisis*, xiv.

27. OpenSecrets, "Lobbying Database," 2014, accessed June 1, 2014, http://www.opensecrets.org/lobby/.

28. Marcus Wolf, Kenneth Haar, and Olivier Hoedeman, *The Fire Power of the Financial Lobby: A Survey of the Size of the Financial Lobby at the EU Level* (April 2014), 15, accessed June 6, 2014, http://corporateeurope.org/sites/default/files/attachments/financial_lobby_report.pdf.

29. Andrew Grice, "Energy Rip-Off: 'Big Six' Firms Too Close to Ministers, Says Ed Miliband." *Guardian*, October 7, 2013, accessed June 22, 2014, http://www.independent.co.uk/news/uk/politics/energy-ripoff-big-six-firms-too-close-to-ministers-says-ed-miliband-8862740.html.

30. International Centre for the Settlement of Investment Disputes, Tecnicas Medioambientales Tecmed S.A. versus the United Mexican States, case no. ARB (AF)/00/2, May 29, 2003, accessed April 1, 2014, http://www.italaw.com/documents/Tecnicas_001.pdf.

31. Government of Canada, Lone Pine Resources Inc. versus the Government of Canada, September 6, 2013, accessed June 1, 2014, http://

www.international.gc.ca/trade-agreements-accords-commerciaux/
assets/pdfs/disp-diff/lone-02.pdf.

32. Bettina Bluemling and Sun-Jin Yun, "Giving Green Teeth to the
Tiger? A Critique of 'Green Growth' in South Korea," in *Green
Growth: Ideology, Political Economy, and the Alternatives*, ed. Gareth
Dale, Manu Mathai, and Jose Puppim de Oliveira, 114–30 (London:
Zed Books, 2016).

33. Cited in Jeong-su Kim, "The Environmental Fallout of the Four
Major Rivers Project," *Hankyoreh*, August 3, 2013, accessed June 8,
2014, http://www.hani.co.kr/arti/english_edition/e_national/598190
.html.

34. Mirowski, *Serious Crisis*, 192.

35. Lee Myung-bak, "Great People"; Lee Myung-bak in Scott A. Snyder, "Assessing the Global Green Growth Institute (GGGI) and
the Sustainability of South Korea's Contribution," *Council on Foreign
Relations*, November 6, 2012, accessed December 1, 2014, http://blogs
.cfr.org/asia/2012/11/06/assessing-the-global-green-growth-institute
-gggi-and-the-sustainability-of-south-koreas-contribution/.

36. World Bank, *Inclusive Green Growth*, 29.

37. Michel Foucault, "Questions of Method," in *The Foucault Effect:
Studies in Governmentality*, ed. Graham Burchell, Colin Gordon, and
Peter Miller (Chicago: University of Chicago Press, 1991), 79.

38. Michel Foucault, "The Subject and Power," *Critical Inquiry* 8, no. 4
(summer 1982): 790.

39. Michel Foucault, *The Birth of Biopolitics: Lectures at the Collège de
France, 1978–1979*, trans. Graham Burchell (New York: Palgrave Macmillan, 2008), 186.

40. James Meadway, "Degrowth and the Roots of Neoclassical Economics," in Dale, Mathai, and Oliveira, *Green Growth*, 97.

41. Friedrich Hayek, *The Road to Serfdom* (Chicago: University of Chicago Press, 1944), 27.

42. Mirowsk, *Serious Crisis*, 8–9.

43. Economic and Social Commission for Asia and the Pacific et al., 89.

44. Ibid.

45. Brad Evans and Julian Reid, *Resilient Life: The Art of Living Dangerously* (Cambridge: Polity Press, 2014).

46. Ibid., 1.
47. Stéphane Hallegatte, Geoffrey Heal, Marianne Fay, and David Treguer, "From Growth to Green Growth: A Framework," World Bank Group Policy Research Working Papers, WPS5872, November 2011.
48. Ibid., 2.
49. Brown, *Undoing the Demos*, 40.

3. GREEN SCARE

1. Core Writing Team, Rajendra K. Pachauri, and Leo Meyer, eds., *Climate Change 2014: Synthesis Report*, accessed March 17, 2017, https://www.ipcc.ch/pdf/assessment-report/ar5/syr/SYR_AR5_FINAL_full.pdf.
2. James Lovelock, *Gaia: A New Look at Life on Earth* (Oxford: Oxford University Press, 1995), 100.
3. Food and Agriculture Organization of the United Nations, *The Water-Energy-Food Nexus: A New Approach in Support of Food Security and Sustainable Agriculture*, June 2014, accessed March 18, 2017, http://www.fao.org/nr/water/docs/FAO_nexus_concept.pdf.
4. "Horn of Africa Food Crisis Remains Dire as Famine Spreads in Somalia," *UN News Centre*, September 5, 2001, accessed December 20, 2014, http://www.un.org/apps/news/story.asp?NewsID=39450#.VJrsosAAvA.
5. Helen Davidson and Nick Evershed, "Australia Bushfires Live: Fears Blue Mountains Fires Will Join Together," *Guardian*, October 21, 2013, accessed December 16, 2014, http://www.theguardian.com/world/2013/oct/21/nsw-fires-residents-evacuate.
6. Debarati Guha-Sapir, Philippe Hoyois, and Regina Below, *Annual Disaster Statistical Review 2014: The Numbers and Trends*, Centre for Research on the Epidemiology of Disasters, September 22, 2014, accessed December 16, 2014, http://www.cred.be/sites/default/files/ADSR_2013.pdf. See also United Nations Central Emergency Response Fund, "Philippines: UN Releases US$25 Million to Fund Emergency Response," November 11, 2013, accessed December 20, 2014, http://www.unocha.org/cerf/resources/top-stories/philippines-un-releases-us25-million-fund-emergency-response.

7. The White House, Office of the Press Secretary, "President Obama Signs Washington Emergency Declaration," March 24, 2014, accessed December 20, 2014, http://www.whitehouse.gov/the-press-office /2014/03/24/president-obama-signs-washington-emergency -declaration.

8. UNEP, UNDP, NATO, OSCE, *Environment and Security: Transforming Risks into Cooperation* (2005), 8, accessed January 2, 2015, http://www.envsec.org/publications/ENVSEC.Transforming%20 risks%20into%20cooperation.%20Central%20Asia.%20Ferghana -Osh-Khujand%20area_English.pdf.

9. Ibid.

10. Ibid.

11. Ibid.,10; emphasis in the original.

12. John Vidal, "Water Supply Key to Outcome of Conflicts in Iraq and Syria, Experts Warn," *Guardian*, July 2, 2014, accessed January 2, 2015, http://www.theguardian.com/environment/2014/jul/02/water-key -conflict-iraq-syria-isis.

13. Carl Schmitt, *Political Theology: Four Chapters on the Concept of Sovereignty*, trans. George Schwab (Cambridge, Mass.: MIT Press, 1985), 13.

14. Ibid., 5.

15. Ibid.

16. Bonnie Honig, "Three Models of Emergency Politics," *Boundary* 41, no. 2 (2014): 46. Honig outlines a tripartite model for understanding emergency politics. She adopts a legalist framework consisting of what she describes as deliberative, promiscuous, and legalist conceptions of emergency states. Critic Elaine Scarry, AIDS activist Douglas Crimp, and lawyer and former assistant U.S. secretary of labor during the Wilson administration, and Louis Freeland Post epitomize each position, respectively.

17. In addition to the two I discuss here as examples of this position is Nomi Claire Lazar, *States of Emergency in Liberal Democracies* (Cambridge: Cambridge University Press, 2009). See also Bonnie Honig, *Emergency Politics: Paradox, Law, Democracy* (Princeton, N.J.: Princeton University Press, 2009).

18. Naomi Klein, *The Shock Doctrine: The Rise of Disaster Capitalism* (New York: Picador, 2007).

19. Giorgio Agamben, *Homo Sacer: Sovereign Power and Bare Life*, trans. Daniel Heller-Roazen (Stanford, Calif.: Stanford University Press, 1998).

20. Agamben writes, "The life caught in the sovereign ban is the life that is originarily sacred—that is, that may be killed but not sacrificed—and, in this sense, the production of bare life is the originary activity of sovereignty. The sacredness of life, which is invoked today as an absolutely fundamental right in opposition to sovereign power, in fact originally expresses precisely both life's subjection to a power over death and life's irreparable exposure in the relation of abandonment" (ibid., 83).

21. Wendy Brown, *Undoing the Demos: Neoliberalism's Stealth Revolution* (New York: Zone Books, 2015), 151.

22. Ibid., 151–52.

23. James F. Jarboe, "Testimony Before the House Resources Committee, Subcommittee on Forests and Forest Health," *Federal Bureau of Investigation*, February 12, 2002, accessed December 22, 2014, http://www.fbi.gov/news/testimony/the-threat-of-eco-terrorism. In 2005 Secretary-General Kofi Annan defined terrorism as that which is "intended to cause death or serious bodily harm to civilians or noncombatants, with the purpose of intimidating a population or compelling a Government or an international organization to do or abstain from doing an act" ("Annan Lays Out Detailed Five-Point UN Strategy to Combat Terrorism," *UN News Centre*, March 10, 2005, accessed December 22, 2015, http://www.un.org/apps/news/story.asp?NewsID=13599&#.VKGu88AAvA).

24. Slavoj Žižek, *The Sublime Object of Ideology* (London: Verso, 1989), 21.

25. Paul Lewis and Rob Evans, "Mark Kennedy: A Journey from Undercover Cop to 'Bona Fide' Activist," *Guardian*, January 10, 2011, accessed December 16, 2014, http://www.theguardian.com/environment/2011/jan/10/mark-kennedy-undercover-cop-activist.

26. Greenpeace, *License to Kill: How Deforestation for Palm Oil Is Driving Sumatran Tigers Toward Extinction*, 2013, accessed January 2, 2015, http://www.greenpeace.org/international/Global/international/publications/forests/2013/LicenceToKill_ENG_LOWRES.pdf.

27. Slavoj Žižek, *The Parallax View* (Cambridge, Mass.: MIT Press, 2006), 10.

28. The logic of my analysis here is obviously indebted to Žižek, *Sublime Object*.

29. Michael Ignatieff, *The Lesser Evil: Political Ethics in an Age of Terror* (Princeton, N.J.: Princeton University Press, 2004).

30. Elaine Scarry, *Thinking in an Emergency* (New York: Norton, 2011), 14.

31. Ibid., 43.

32. Ibid., 61.

33. Scarry (ibid., 17–18, 41, 53, 57) provides a useful comparison between the Swiss and United States shelters system to demonstrate the equality of survival principle. In the United States, the "very individuals who had the nuclear arsenal at their disposal . . . continued to spend billions of dollars on an extensive shelter system for themselves in Mount Weather in the Blue Ridge Mountains of Virginia"; the Swiss shelter system is designed for the benefit of all inhabitants of Switzerland.

34. Martin Gilens and Benjamin I. Page, "Testing Theories of American Politics: Elites, Interest Groups, and Average Citizens," *Perspectives on Politics* 12, no. 3 (summer 2014): 564–81.

35. Retort, *Afflicted Powers: Capital and Spectacle in a New Age of War* (London: Verso, 2005), 72.

36. Rob Evans and Paul Lewis, "Revealed: How Energy Firms Spy On Environmental Activists," *Guardian*, February 14, 2011, accessed December 21, 2014, http://www.theguardian.com/environment/2011/feb/14/energy-firms-activists-intelligence-gathering.

37. Honig, "Three Models," 48.

38. Adam Bandt, "Had We but World Enough and Time," *Australian Feminist Law Journal* 31, no. 1 (December 2009): 18.

39. Ibid.

40. Bandt as discussed in Honig, "Three Models," 49.

41. Slavoj Žižek, *Looking Awry: An Introduction to Jacques Lacan Through Popular Culture* (Cambridge, Mass.: MIT Press, 1992), 14.

42. Gilles Deleuze and Félix Guattari, *A Thousand Plateaus: Capitalism and Schizophrenia*, trans. Brian Massumi (Minneapolis: University of Minnesota Press, 1987), 214.

43. Ibid., 215.

44. Žižek, *Looking Awry*, 6.

45. Ibid., 7.

46. Ibid., 8.

47. Slavoj Žižek, *Violence: Six Sideways Reflections* (New York: Picador, 2008), 9–14.

48. Walter Benjamin, "On the Concept of History," in *Walter Benjamin: Selected Writings, Volume 4, 1938–1940*, ed. Howard Eiland and Michael W. Jennings (Cambridge, Mass.: Belknap Press, 2006), 392.

49. "The truth of the matter is that *social production is purely and simply desiring-production itself under determinate conditions*. We maintain that the social field is immediately invested by desire, that it is the historically determined product of desire, and that libido has no need of any mediation or sublimation, any psychic operation, any transformation, in order to invade and invest the productive forces and the relations of production. *There is only desire and the social, and nothing else*" (Gilles Deleuze and Félix Guattari, *Anti-Oedipus: Capitalism and Schizophrenia*, trans. Robert Hurley, Mark Seem, and Helen R. Lane [London: Continuum, 2004], 31; emphasis in the original).

50. Brad Evans and Julian Reid, *Resilient Life: The Art of Living Dangerously* (Cambridge: Polity Press, 2014), 171.

51. Ibid.

52. Here I am using the concept of the multitude developed in Michael Hardt and Antonio Negri, *Multitude: War and Democracy in the Age of Empire* (New York: Penguin, 2004).

4. FASCIST EARTH

1. Wendy Brown, *Undoing the Demos: Neoliberalism's Stealth Revolution* (New York: Zone Books, 2015).

2. Kenneth Surin, *Freedom Not Yet: Liberation and the Next World Order* (Durham, N.C.: Duke University Press, 2009).

3. Ibid., 30.

4. Brown, *Undoing the Demos*, 87.

5. Ibid., 110.

6. Murray Bookchin, *Social Ecology and Communalism* (Oakland, Calif.: AK Press, 2006), 19–20.

7. Ibid.

8. Ibid.; Murray Bookchin, *The Ecology of Freedom: The Emergence and Dissolution of Hierarchy* (Palo Alto, Calif.: Cheshire Books, 1982).

9. Murray Bookchin, *The Next Revolution: Popular Assemblies and the Promise of Direct Democracy*, ed. Debbie Bookchin and Blair Taylor (London: Verso, 2015).

10. David Fairhall, *Common Ground: The Story of Greenham* (London: St. Martin's Press, 2000), 6–9.

11. Christina Welch, "The Spirituality of, and at, Greenham Common Peace Camp," *Feminist Theology* 18, no. 2 (2010): 232; Sarah Hipperson, *Greenham Common Women's Peace Camp*, accessed July 20, 2016, http://www.greenhamwpc.org.uk/.

12. Fairhall, *Common Ground*, 8.

13. Welch, "Spirituality," 233–34.

14. Paul Brown, "Protest by CND Stretches 14 Miles," *Guardian*, April 2, 1983, accessed July 20, 2016, https://www.theguardian.com/fromthearchive/story/0,,1866956,00.html.

15. Sophie Mayer, "The Legend of Greenham Common Women's Peace Camp," *Transformation* February 2, 2016, accessed July 20, 2016, https://www.opendemocracy.net/transformation/sophie-mayer/bringing-home-legend-of-greenham-common-womens-peace-camp.

16. Fairhall, *Common Ground*, 9.

17. Ibid.

18. Judith Butler and Athena Athanasiou, *Dispossession: The Performative in the Political* (Cambridge: Polity Press, 2013), 14.

19. David Logie, "On the Frontlines of the Refugee Crisis," *Greenpeace*, April 3, 2016, accessed July 17, 2015, http://www.greenpeace.org/usa/greenpeace-european-refugee-crisis-me/.

20. Michael Brune, "Sierra Club Statement on Police Killings of Alton Sterling and Philando Castile," *Sierra Club*, July 7, 2016, accessed July 17, 2016, http://content.sierraclub.org/press-releases/2016/07/sierra-club-statement-police-killings-alton-sterling-and-philando-castile.

21. Fred Krupp, "Statement of Environmental Defense Fund President Fred Krupp on Recent Events in Baton Rouge, Minneapolis, and Dallas," *Environmental Defense Fund*, July 8, 2016, accessed July 17, 2016, https://www.edf.org/media/statement-environmental-defense-fund-president-fred-krupp-recent-events-baton-rouge.

22. John S. Dryzek, *The Politics of the Earth*, 2nd ed. (Oxford: Oxford University Press, 2005), 8.

23. Ibid., 10.

24. National-Anarchist Movement, "Part 7: The Green Replenishment," September 18, 2010, accessed July 20, 2016, http://www.national -anarchist.net/2010/09/part-7-green-replenishment.html.

25. Ibid.

26. National-Anarchist Movement, "Part 9: Defence," September 18, 2010, accessed July 21, 2016, http://www.national-anarchist.net/2010 /09/part-9-defence.html.

27. National-Anarchist Movement, "Part 5: Racial Separatism or Mixed Tribes?" September 18, 2010, accessed July 21, 2016, http://www .national-anarchist.net/2010/09/part-5-racial-separatism.html.

28. Ibid.

29. The Left was so enraged by the Right's infiltrating its protest that police were forced to protect right-wing activists from the Left's assault. See Spencer Sunshine, "Rebranding Fascism: National-Anarchists," *Public Eye* 23, no. 4 (winter 2008), accessed July 21, 2016, http://www.publiceye.org/magazine/v23n4/rebranding_fascism .html.

30. Black Sun Invictus, "Interview with Troy Southgate," *National-Anarchist Movement*, May 29, 2012, accessed July 20, 2016, http://www .national-anarchist.net/2012/05/interview-with-troy-southgate-from .html.

31. Graham D. Macklin, "Co-opting the Counter Culture: Troy Southgate and the National Revolutionary Faction," *Patterns of Prejudice* 39, no. 3 (2005): 306.

32. Ibid., 307–8.

33. Miron Fyodorov, "Interview with Troy Southgate for *Kinovar* (Russia)," *Euro-Synergies*, February 5, 2008, accessed July 17, 2016, http:// www.thephora.net/forum/archive/index.php/t-33981.html.

34. Macklin, "Co-opting the Counter Culture," 310.

35. Ibid., 313.

36. Troy Southgate, "Transcending the Beyond: Third Position to National-Anarchism," *Pravda*, January 17, 2002, accessed July 21, 2016, http://www.pravdareport.com/news/russia/17-01-2002/25940-0/.

37. For more on the racist framing used in the argument linking population with environmental degradation, see Adrian Parr, *The Wrath of
Capital: Neoliberalism and Climate Change Politics* (New York: Columbia University Press, 2013), 39–51.

38. Heidi Beirich, *Greenwash: Nativists, Environmentalism, and the
Hypocrisy of Hate*, ed. Mark Potok, Southern Poverty Law Center
(July 2010), 7, accessed July 21, 2016, https://www.splcenter.org/sites/
default/files/d6_legacy_files/downloads/publication/Greenwash
.pdf. From 1975 to 1977 John Tanton was the president of the organization Zero Population Growth.

39. Ibid., 16.

40. Ibid.

41. Betsy Hartmann, "The Greening of Hate: An Environmentalist's
Essay," in Beirich, *Greenwash*.

42. Frank N. Egerton, *Roots of Ecology: Antiquity to Haeckel* (Berkeley:
University of California Press, 2012), 2.

43. Raymond H. Dominick, *The Environmental Movement in Germany:
Prophets and Pioneers, 1871–1971* (Bloomington: Indiana University
Press, 1992), 22.

44. Egerton, *Roots of Ecology*, 74; Peter Staudenmaier, "Fascist Ecology:
The 'Green Wing' of the Nazi Party and Its Historical Antecedents,"
in *Ecofascism: Lessons from the German Experience*, by Janet Biehl and
Peter Staudenmaier (Oakland, Calif.: AK Press, 1995), 7.

45. Arnold Arluke and Clinton Sanders, *Regarding Animals* (Philadelphia: Temple University Press, 2012), 140–42.

46. Daniel Gasman, *The Scientific Origins of National Socialism* (London:
Transaction, 2007), xli.

47. Wilhelm Heinrich Riehl, "Field and Forest" (1853), trans. Frances H.
King, in *The German Classics: Volume VIII* (1913), ed. Kuno Francke,
416, accessed July 22, 2016, http://www.unz.org/Pub/FranckeKuno
-1913v08-00410.

48. Cited in Robert G. Lee and Sabine Wilke, "Forest as *Volk: Ewiger
Wald* and the Religion of Nature in the Third Reich," *Journal of Social
and Ecological Boundaries* 1, no. 1 (spring 2005): 22.

49. Fritz Nova, *Alfred Rosenberg: Nazi Theorist of the Holocaust* (New
York: Hippocrene, 1986).

50. Gasman, *Scientific Origins*; Biehl and Staudenmaier, *Ecofascism*; George L. Mosse, *The Crisis of German Ideology: Intellectual Origins of the Third Reich* (New York: Grosset and Dunlap, 1964); Petteri Pietikäinen, "The Volk and Its Unconscious: Jung, Hauer, and the German Revolution," *Journal of Contemporary History* 35, no. 4 (October 2000): 523–39.

51. Jürgen Matthäus and Frank Bajohr, *The Political Diary of Alfred Rosenberg and the Onset of the Holocaust* (Lanham, Md.: Rowman and Littlefield, 2015), 403.

52. Cited in Pietikäinen, "Volk and Its Unconscious," 523.

53. Ibid.; emphasis added.

54. Adolf Hitler, *Mein Kampf*, trans. James Murphy (1924), 379, accessed July 22, 2016, http://www.greatwar.nl/books/meinkampf/meinkampf .pdf.

55. Mosse, *Crisis of German Ideology*, 267.

56. Hitler, *Mein Kampf*, 316–17.

57. Robert A. Pois, *National Socialism and the Religion of Nature* (New York: St. Martin's Press, 1986).

58. Frank Uekoetter, *The Green and the Brown: A History of Conservation in Nazi Germany* (Cambridge: Cambridge University Press, 2006), 1.

59. Darré expressed his attachment to the ideology of blood and soil through a variety of public programs he instituted. See Anna Bramwell, *Blood and Soil: Richard Walther Darré and Hitler's "Green Party."* (Bourne End, Buckinghamshire, U.K.: Kensal Press, 1984), 4.

60. Arluke and Sanders, *Regarding Animals*, 133–37.

61. Cited ibid., 141–42.

62. Nazi propaganda in the magazine *Neugeist: Die weiße Fahne*, cited ibid., 148.

63. Wilhelm Reich, *The Mass Psychology of Fascism* (New York: Farrar, Straus and Giroux, 1970), 216–17.

64. Ibid., 232.

65. Surin, *Freedom Not Yet*, 53.

66. Cited in Eric Kurlander, "Hitler's Monsters: The Occult Roots of Nazism and the Emergence of the Nazi 'Supernatural Imaginary,'" *German History* 30, no. 4 (2012): 528.

67. Surin, *Freedom Not Yet*, 248.

68. Gianni Vattimo and Santiago Zabala, *Hermeneutic Communism: From Heidegger to Marx* (New York: Columbia University Press, 2011).

69. Ibid.

70. Ibid.

71. Weakening the historical significance of communism entails deepening the influence of local actors in state politics and interpreting communism anew by decentralizing the static and the heavy-handed top-down bureaucratic functioning of Soviet-style communism. Ibid., 42.

72. Ibid., 43.

73. Douglas Torgerson, *The Promise of Green Politics: Environmentalism and the Public Sphere* (Durham, N.C.: Duke University Press, 1999), 3.

74. Andrew Dobson, *Green Political Thought*, 2nd ed. (London: Routledge, 1995), 32–35.

75. Adrian Parr and Natasha Leonard, "Our Crime Against the Planet and Ourselves," *New York Times*, May 18, 2016, accessed July 1, 2016, http://www.nytimes.com/2016/05/18/opinion/our-crime-against-the -planet-and-ourselves.html?_r=0.

76. Michel Foucault and Gilles Deleuze wrote, "A theory is exactly like a box of tools. It has nothing to do with the signifier. It must be useful. It must function. And not for itself. If no one uses it, beginning with the theoretician himself (who then ceases to be a theoretician), then the theory is worthless or the moment is inappropriate" ("Intellectuals and Power," in *Language, Counter-Memory, Practice: Selected Essays and Interviews*, by Michel Foucault, trans. Donald F. Bouchard and Sherry Simon [Ithaca, N.Y.: Cornell University Press, 1977], 208).

5. COMMONISM

1. The architecture of Detroit spoke volumes of the city's prosperous past. Detroit has some incredible examples of Art Deco skyscrapers such as the Book-Cadillac Hotel (1924), designed by Louis Kamper, along with several magnificent leftovers from the 1880s Gilded Age architects, such as Harry J. Rill and Gordon Lloyd, not to forget

wonderful examples of American neoclassical architecture like the Cadillac Place (1923), designed by Albert Kahn for General Motors.

2. Michael Hardt and Antonio Negri, *Commonwealth* (Cambridge, Mass.: Belknap Press, 2009), 252.

3. Pittsburgh is an example of one old industrial center adapting to the changing economic landscape of neoliberal doctrine. When its steel industry collapsed in the 1980s, the city diversified its economy to include the energy, finance, and health industries, along with proactively recruiting young people to the city by revitalizing the livability of the downtown.

4. Philipp Oswalt and Tim Rieniets, *Atlas of Shrinking Cities* (Stuttgart: Hatje Cantz, 2006), 148–50.

5. Jeffrey S. Zax and John F. Kain, "Moving to the Suburbs: Do Relocating Companies Leave Their Black Employees Behind?" *Journal of Labor Economics* 14, no. 3 (1996): 472–504.

6. Ted Mouw, "Job Relocation and the Racial Gap in Unemployment in Detroit and Chicago, 1980 to 1990," *American Sociological Review* 65, no. 5 (2000): 736.

7. Ibid., 738.

8. Ibid.

9. Ibid., 740.

10. John Haaga, Richard Scott, and Jennifer Hawes-Dawson, *Drug Use in the Detroit Metropolitan Area: Problems, Programs, and Policy Options* (Santa Monica, Calif.: RAND, 1992), v, accessed December 20, 2014, http://www.rand.org/content/dam/rand/pubs/reports/2009/R4085 .pdf.

11. Philipp Misselwitz, "Shrinking Cities: Manchester/Liverpool," Working Papers (March 2004), 115–18, accessed March 1, 2013, http: //www.shrinkingcities.com/fileadmin/shrink/downloads/pdfs/WP -II_Manchester_Liverpool.pdf.

12. Oswalt and Rieniets, *Atlas of Shrinking Cities*, 148–49.

13. Misselwitz, "Shrinking Cities," 115–18.

14. Nouriel Roubini and Stephen Mihm, *Crisis Economics: A Crash Course in the Future of Finance* (New York: Penguin, 2010).

15. Joseph E. Stiglitz, *Free Fall: America, Free Markets, and the Sinking of the World Economy* (New York: Norton, 2010).

16. Roubini and Mihm, *Crisis Economics*.

17. Sharon Smyth and Charles Penty, "Spain Foreclosures Spread to Once Wealthy," *Bloomberg*, October 9, 2012, accessed December 7, 2012, http://www.bloomberg.com/news/articles/2012-10-08/spain -foreclosures-spread-to-once-wealthy-mortgages.

18. Hilary Osborne, "Repossessions at Highest Level Since 1995," *Guardian*, February 11, 2010, accessed December 21, 2012, http://www .theguardian.com/money/2010/feb/11/home-repossessions-highest -level-1995.

19. British Broadcasting Corporation, "Europe in Housing Market 'Agony,' Says Rics," *BBC Business News*, February 28, 2012, accessed December 21, 2012, http://www.bbc.co.uk/news/business-17191691.

20. Les Christie, "Foreclosures up a Record 81% in 2008," *CNN Money*, January 15, 2009, accessed December 22, 2012, http://money.cnn.com /2009/01/15/real_estate/millions_in_foreclosure/.

21. RealtyTrac, "Foreclosure Activity Increases 4 Percent in July," August 12, 2010, Accessed March 3, 2011, http://www.realtytrac.com/content/ press-releases/foreclosure-activity-increases-4-percent-in-july-5946.

22. Karl Marx, *Capital: Volume 1*, trans. Ben Fowkes (London: Penguin, 1990), 166.

23. David Harvey, *Rebel Cities: From the Right to the City to Urban Revolution* (London: Verso, 2012), 5.

24. Ibid.

25. Henri Lefebvre, *The Urban Revolution*, trans. Robert Bononno (Minneapolis: University of Minnesota Press, 2003).

26. Ibid., 57.

27. Andrea Kahn, "Defining Urban Sites," in *Site Matters: Designs, Concepts, Histories, and Strategies*, ed. Carol Burns and Andrea Kahn (New York: Routledge, 2005), 286.

28. Andy Merrifield, "The Urban Question under Planetary Urbanization," *International Journal of Urban and Regional Research* 37, no. 3 (May 2013): 912.

29. Neil Brenner, "Theses on Urbanization," *Public Culture* 25, no. 1 (2013): 91.

30. Ibid.; Merrifield, "Urban Question."

31. David Harvey *A Brief History of Neoliberalism* (Oxford: Oxford University Press, 2005), 7.

32. Ibid.

33. David Harvey, *The Enigma of Capital* (Oxford: Oxford University Press, 2010), 204–5.

34. Merrifield, "Urban Question," 914.

35. Ibid. See also David Harvey, *Spaces of Global Capitalism: A Theory of Uneven Geographical Development* (London: Verso, 2006).

36. Mike Davis, *Planet of Slums* (London: Verso, 2006), 3.

37. Ibid.; see also Shlomo Angel, *Planet of Cities* (Cambridge, Mass.: Lincoln Institute of Land Policy, 2012), and Robert Neuwirth, *Shadow Cities: A Billion Squatters, a New Urban World* (New York: Routledge 2006).

38. Pierre Bourdieu is typically associated with having inaugurated the concept of social capital, positing a circular argument that describes how the powerful and privileged secure their power through the social networks they have with other powerful and privileged individuals; see Pierre Bourdieu, "The Forms of Capital," in *Handbook of Theory and Research for the Sociology of Education*, ed. John Richardson, 241–58 (New York: Greenwood Press, 1986). James Coleman picks up on Bourdieu's project, expanding the concept of social capital to include social relations that provide nonelite actors with resources that can counterbalance the inequities they encounter and endure. Coleman's focus on social capital arising from neighborhood groups and kinship ties has a distinctively conservative and traditionalist ring to it; see James Coleman, "Social Capital in the Creation of Human Capital," *American Journal of Sociology* 94, no. 51 (1988): 95–120. Similarly, Robert Putnam laments over what he perceives to be a deficit of civic engagement in contemporary life. The reason he provides is the demise of social capital. The solution, he says, comes from revitalizing social networks, norms, relations of trust, and cooperation among individuals pursuing shared objectives; see Robert D. Putnam, *Making Democracy Work: Civic Traditions in Modern Italy* (Princeton, N.J.: Princeton University Press, 1993); Robert Putnam, "Bowling Alone: America's Declining Social Capital," *Journal of Democracy* 6, no. 1 (1995): 65–78. All social capital theorists maintain that a social network is productive of capital.

39. Sheila Foster, "The City as an Ecological Space: Social Capital and Urban Land Use," *Notre Dame Law Review* 82, no. 2 (2006): 545.

40. Coleman, "Social Capital"; Putnam, *Making Democracy Work*; Putnam, "Bowling Alone,"; Robert Putnam, *Bowling Alone: The Collapse and Revival of American Community* (New York: Simon and Schuster, 2000).

41. Bobbi Dempsey and Todd Beitler, *The Complete Idiot's Guide to Buying Foreclosures*, 2nd ed. (New York: Penguin, 2005), xix.

42. Vanessa Wong, "Goodbye, Renters Market: Rents Grow as Demand Increases," *Bloomberg Business Week*, March 21, 2011, accessed December 12, 2012, http://www.realestate.msn.com/article.aspx?cp-documentid =28001698.

43. Solomon Moore, "As Program Moves Poor to Suburbs, Tensions Follow," *New York Times*, August 8, 2008, accessed March 16, 2011, http://www.nytimes.com/2008/08/09/us/09housing.html?pagewanted =all&_r=0.

44. Alan Berube, Audrey Singer, and William Frey, *The State of Metropolitan America*, Brookings Institution, 2010, accessed March 21, 2011, http://www.brookings.edu/~/media/research/files/reports/2010/5 /09%20metro%20america/metro_america_report.pdf.

45. Nico Larco, "Untapped Density: Site Design and the Proliferation of Suburban Multifamily Housing," *Journal of Urbanism: International Research on Placemaking and Urban Sustainability* 2, no. 2 (2009): 167–86.

46. Foster, "City as an Ecological Space," 546.

47. Ibid., 574.

48. Ibid., 573.

49. Steve Pardo, "140 Acres in Detroit Sold to Grow Trees," *Detroit News*, December, 12, 2012, accessed December 14, 2012, http://www .detroitnews.com/article/20121212/METRO01/212120340.

50. Garrett Hardin, "The Tragedy of the Commons," *Science* 162, no. 3859 (1968): 1243–48.

51. Ostrom shared the Nobel Prize with Oliver Williamson.

52. Elinor Ostrom, *Governing the Commons: The Evolution of Institutions for Collective Action* (Cambridge: Cambridge University Press, 1990).

53. Sinan Koont, "The Urban Agriculture of Havana," *Monthly Review* 60, no. 8 (2009), accessed March 1, 2013, http://monthlyreview.org /2009/01/01/the-urban-agriculture-of-havana/.

54. Ibid.

55. Ibid.

56. Embassy of the Bolivarian Republic of Venezuela to the U.S., *Fact Sheet: Urban Agriculture in Venezuela*, August 2012, accessed March 1, 2013, http://venezuela-us.org/live/wp-content/uploads/2009/08/08.20 .2012-Urban-Agriculture-ENG1.pdf.

57. Cited ibid., 1.

58. Ibid.

59. Ibid.

60. Hardt and Negri, *Commonwealth*, viii.

61. Public space is an area that can be openly accessed by any person regardless of race, gender, class, or ethnicity. Public space is a bounded area defined largely in opposition to private property. A public space is also one where civic life takes place, in the sense that it facilitates the processes of participatory democracy whereby people express their opinions publicly and communicate common concerns. That said, the systems of securitization interwoven throughout public space may not restrict access to public areas, but they do discipline civic life through mechanisms of surveillance and other socially coercive mechanisms such as the norms of "civil" conduct that often exclude nonnormative forms of behavior, such as the hostility LGBTQ youth, the homeless, and street vendors encounter in public space. What this means is that the form and content of public space are restricted.

62. Hardt and Negri, *Commonwealth*.

63. Ibid., 15–16.

64. Ibid., 250.

65. Gilles Deleuze and Félix Guattari, *A Thousand Plateaus: Capitalism and Schizophrenia*, trans. Brian Massumi (Minneapolis: University of Minnesota Press, 1987), 161.

6. WELCOME TO THE DARK SIDE OF DIGNITY AND DEVELOPMENT

1. Frantz Fanon, *Black Skin, White Masks*, trans. Richard Philcox (New York: Grove Press, 2008), 170.

2. Ananya Roy, "Encountering Poverty," in *Encountering Poverty: Thinking and Acting in an Unequal World*, by Ananya Roy, Genevieve Negrón-Gonzales, Kweku Opoku-Agyemang, and Clare Talwalker (Berkeley: University of California Press, 2016), 31.

3. Ibid., 28–29.

4. United Nations Environment Program, *Kenya: Atlas of Our Changing Environment*, 2009, accessed March 17, 2017, http://staging.unep .org/pdf/Kenya_Atlas_Full_EN_72dpi.pdf.

5. Amnesty International, *Kenya: The Unseen Majority; Nairobi's Two Million Slum Dwellers* (2009), 1, accessed February 3, 2016, https:// www.amnesty.nl/sites/default/files/public/rap_kenia_the_unseen_ majority.pdf.

6. Millennium Development Goals Indicators, accessed February 15, 2016, http://mdgs.un.org/unsd/mdg/Metadata.aspx?IndicatorId=0& SeriesId=711.

7. Co-operative Housing International, "About Kenya," accessed February 14, 2016, http://www.housinginternational.coop/co-ops/kenya.

8. Benjamin Chavis quoted in Richard Lazarus, "Environmental Racism! That's What It Is," *University of Illinois Law Review* 2000, no. 1 (2000): 257.

9. Jason Corburn and Chantal Hildebrand, "Slum Sanitation and the Social Determinant of Women's Health in Nairobi, Kenya," *Journal of Environmental and Public Health* 2015 (2015): 1–6.

10. Human subjects protected under University of Cincinnati's IRB protocol no. 2013-5966.

11. Ban Ki Moon, *The Road to Dignity by 2030: Ending Poverty, Transforming All Lives and Protecting the Planet* (December 2014), 6, accessed December 26, 2015, http://www.un.org/ga/search/view_doc.asp ?symbol=A/69/700&Lang=E.

12. Michael Rosen, *Dignity* (Cambridge, Mass.: Harvard University Press, 2012), 11–14; Eduardo Mendieta, "The Legal Orthopedia of

Human Dignity: Thinking with Axel Honneth," *Philosophy and Social Criticism* 40, no. 8 (2014): 803.

13. Cited in Rosen, *Dignity*, 12.

14. Michael Rosen identifies this view of dignity with Francis Bacon, Gelasius, and Aquinas; ibid., 15–18.

15. Ibid., 16.

16. Ibid., 17.

17. Thomas Aquinas, *Summa Theologiae*, vol. 38, *Injustice*, ed. Marcus Lefé-bure (Cambridge: Cambridge University Press, 2006), 25.

18. Immanuel Kant, *Groundwork of the Metaphysics of Morals*, ed. Mary Gregor and Jens Timmermann, rev. ed. (Cambridge: Cambridge University Press, 2012), 47.

19. The idea of the *imago dei* reappeared during the civil rights movement in the United States and was integral to Martin Luther King Jr.'s campaign: "The 'image of God' is the idea that all men have something within them that God injected. Not that they have substantial unity with God, but that every man has a capacity to have fellowship with God. And this gives him a uniqueness, it gives him worth, it gives him dignity . . . there are no gradations in the image of God . . . precisely because every man is made in the image of God . . . We will know one day that God made us to live together as brothers and to respect the dignity and worth of every man."

King's activism specifically drew upon a Christian view of dignity. That is, as God's children, human beings share an equal value, regardless of the color of their skin. Having been created in the image of God, God is the origin of human value and dignity, and that value is infinite. See Martin Luther King Jr., "The American Dream," delivered at Ebenezer Baptist Church, Atlanta, Georgia, July 4, 1965, accessed January 31, 2016, http://kingencyclopedia .stanford.edu/encyclopedia/documentsentry/doc_the_american_ dream/.

20. Kant, *Groundwork*, 46.

21. United States Conference of Catholic Bishops, "Renewing the Earth," November 14, 1991, accessed February 2, 2016, http://www .usccb.org/issues-and-action/human-life-and-dignity/environment/ renewing-the-earth.cfm.

22. George Kateb, *Human Dignity* (Cambridge, Mass.: Belknap Press, 2011), 173.

23. Ibid., 144.

24. Ibid., 134.

25. Ibid., 24.

26. Rosi Braidotti, *The Posthuman* (Cambridge: Polity Press, 2013), 13.

27. Ibid., 15.

28. Jim Yong Kim, "Speech by World Bank Group President Jim Yong Kim at the Migration and the Global Development Agenda," *World Bank*, December 9, 2015, accessed December 26, 2015, http://www.worldbank.org/en/news/speech/2015/12/09/speech-by-world-bank-group-president-jim-yong-kim-at-the-migration-and-the-global-development-agenda.

29. "The Universal Declaration of Human Rights (UDHR) is a milestone document in the history of human rights. Drafted by representatives with different legal and cultural backgrounds from all regions of the world, the Declaration was proclaimed by the United Nations General Assembly in Paris on 10 December 1948 (General Assembly resolution 217 A) as a common standard of achievements for all peoples and all nations. It sets out, for the first time, fundamental human rights to be universally protected" (United Nations, "Universal Declaration of Human Rights," accessed January 15, 2016, http://www.un.org/en/universal-declaration-human-rights/).

30. World Bank, "About the World Bank," accessed February 1, 2016, http://www.worldbank.org/en/about.

31. Bretton Woods Project, "Analysis of World Bank Voting Reforms," briefing, April 30, 2010, accessed February 1, 2016, http://www.brettonwoodsproject.org/2010/04/art-566281/.

32. Gustavo Esteva, Salvatore Babones, and Philipp Babicky, *The Future of Development: A Radical Manifesto* (Bristol, U.K.: Policy Press, 2013).

33. United Nations, "A Life of Dignity for All: Accelerating Progress towards the Millennium Development Goals and Advancing the United Nations Development Agenda beyond 2015: Report of the Secretary-General, July 26, 2013, accessed January 21, 2016, http://www.un.org/millenniumgoals/pdf/A%20Life%20of%20Dignity%20for%20All.pdf.

34. United Nations, "We Can End Poverty: Millennium Development Goals and Beyond 2015," *UN Millennium Goals*, accessed January 21, 2016, http://www.un.org/millenniumgoals/environ.shtml.

35. Arturo Escobar, *Encountering Development: The Making and Unmaking of the Third World* (Princeton, N.J.: Princeton University Press, 2012), 213.

36. Here I align myself with the double logic of dignity explicated in Scott Cutler Shershow, *Deconstructing Dignity: A Critique of the Right-to-Die Debate* (Chicago: University of Chicago Press, 2014).

37. Cited in Sean Homer, *Fredric Jameson: Marxism, Hermeneutics, Postmodernism* (Cambridge: Cambridge University Press, 1998), 15.

38. World Bank, "Extreme Poverty Rates Continue to Fall," *World Bank*, June 2, 2010, accessed February 1, 2016, http://blogs.worldbank.org/opendata/extreme-poverty-rates-continue-fall.

39. Raakel Syräjnen, *UN-HABITAT and the Kenya Slum Upgrading Programme: Strategy Document* (UN-HABITAT, 2008), 23, accessed February 11, 2016, https://unhabitat.org/books/un-habitat-and-the-kenya-slum-upgrading-programme-strategy-document/.

40. Katherine N. Rankin, "Social Capital, Microfinance, and the Politics of Development," *Feminist Economics* 8, no. 1 (2002): 1–24.

41. Marie Huchzermeyer, "Slum Upgrading in Nairobi within the Housing and Basic Services Market," *Journal of Asian and African Studies* 43, no. 1 (2008): 19–39.

42. Roy, "Encountering Poverty," 31.

43. Fredric Jameson, *Marxism and Form: Twentieth-Century Dialectical Theories of Literature* (Princeton, N.J.: Princeton University Press, 1971).

44. Adrian Parr, *Hijacking Sustainability* (Cambridge, Mass.: MIT Press, 2009), 141.

45. Edward Said, "On Dignity and Solidarity" (lecture, Washington, D.C., June 15, 2003), *Democracy Now*, accessed February 11, 2016, http://www.democracynow.org/2003/10/20/on_dignity_and_solidarity_scholar_activist.

46. Slavoj Žižek, *Trouble in Paradise: From the End of History to the End of Capitalism* (London: Allen Lane, 2014), 146.

47. Michael Hardt and Antonio Negri, *Empire* (Cambridge, Mass.: Harvard University Press, 2000), 314–19.

7. URBAN CLEAR-CUTTING

1. Cited in Aldo Rossi, *The Architecture of the City*, trans. Diane Ghirardo and Joan Ockman (Cambridge, Mass.: MIT Press, 1982), 107.

2. There is a whole movement committed to greening the military, an initiative I have elsewhere criticized as an oxymoron. That is to say, it is nonsensical to create weaponry that carries a minimal ecological footprint when the weapon is used for killing and injury. The whole point of the military is to exert control through the use of violence; this is the antithesis of sustainability. See Adrian Parr, *Hijacking Sustainability* (Cambridge, Mass.: MIT Press, 2009), 79–92.

3. David Orr, *The Nature of Design: Ecology, Culture, and Human Intention* (Oxford: Oxford University Press, 2002); Alex Steffen, *Carbon Zero: Imagining Cities That Can Save the Planet* (Mountain View, Calif.: Creative Commons, 2012); David Owen, *Green Metropolis: Why Living Smaller, Living Closer, and Driving Less Are the Keys to Sustainability* (New York: Riverhead Books, 2009); Maf Smith, John Whitelegg, and Nick Williams, *Greening the Built Environment* (London: Routledge, 1998); Sadhu Aufochs Johnston, Steven Nicholas, and Julia Parzen, *The Guide to Greening Cities* (Washington, D.C.: Island Press, 2013).

4. Adam Withnall, "Global Peace Index 2016: There Are Now Only 10 Countries in the World That Are Actually Free from Conflict," *Independent*, June 8, 2016, accessed August 11, 2016, http://www.independent.co.uk/news/world/politics/global-peace-index-2016-there-are-now-only-10-countries-in-the-world-that-are-not-at-war-a7069816.html.

5. UNHCR, "Worldwide Displacement Hits All-Time High as War and Persecution Increase," June 18, 2015, accessed August 5, 2016, http://www.unhcr.org/en-us/news/latest/2015/6/558193896/worldwide-displacement-hits-all-time-high-war-persecution-increase.html.

6. Alain Badiou has situated this dilemma between the U.S. Constitution and the French Revolution: "The real difference with Hannah Arendt should, rather, be located in her definition of politics itself. For Arendt, politics concerns 'living together', the regulation of being together as a republic, or as public space. It's not an adequate definition. It reduces politics to the sole instance of judgement, and eventually to

opinion, rather than recognizing that the essence of politics concerns thought and action, as connected through the practical consequences of a prescription. For any one prescription is opposed to others. There can be no homogeneous public space other than that of consensus— the consensus we are all familiar with, the consensus of *la pensée unique* [i.e. global liberalism] . . . On the one hand, the constitutional creation of a complex ramified public space, elaborated in detail . . . On the other hand, something sequential, something more antagonistic, and more principled. I stand resolutely for the second option" (*Ethics: An Essay on the Understanding of Evil*, trans. Peter Hallward [London: Verso 2001], 115–16).

7. Alain Badiou, *Being and Event*, trans. Oliver Feltham (London: Continuum, 2007); Badiou, *Ethics*, 40–57.

8. United Nations, *Urban Millennium* (June 6–8, 2001), 6, accessed August 8, 2016, http://www.un.org/ga/Istanbul+5/booklet4.pdf.

9. United Nations, "World's Population Increasingly Urban with More Than Half Living in Urban Areas," July 10, 2014, accessed August 8, 2016, http://www.un.org/en/development/desa/news/population/world-urbanization-prospects-2014.html.

10. The agglomeration index (density, distance, and division) is a new method for measuring urban concentration. It aims to provide a definition that is globally consistent and can be used for cross-country comparisons. See World Bank, *World Development Report 2009: Reshaping Economic Geography* (Washington, D.C.: World Bank, 2008).

11. United States Census Bureau, "2010 Census Urban Area FAQs: Urban-Rural Classification Program," accessed August 8, 2016, https://www.census.gov/geo/reference/ua/uafaq.html.

12. Eurostat, "Glossary: Predominantly Urban Region," *Statistics Explained*, accessed August 8, 2016, http://ec.europa.eu/eurostat/statistics-explained/index.php/Glossary:Predominantly_urban_region.

13. Eurostat, "Glossary: Degree of Urbanisation," *Statistics Explained*, accessed August 8, 2016, http://ec.europa.eu/eurostat/statistics-explained/index.php/Glossary:Degree_of_urbanisation.

14. Government of India, Ministry of Home Affairs, "Census of India 2011: Provisional Population Totals; Urban Agglomerations and Cities," accessed August 8, 2016, http://censusindia.gov.in/2011-prov-results/paper2/data_files/India2/1.%20Data%20Highlight.pdf.

15. Leo Hollis, *Cities Are Good for You: The Genius of the Metropolis* (New York: Bloomsbury, 2013), 10.

16. Edward Glaeser, *Triumph of the City: How Urban Spaces Make Us Human* (London: Macmillan, 2011), 6.

17. Jane Jacobs, *The Death and Life of Great American Cities* (New York: Random House, 1961).

18. Hollis, *Cities*.

19. Ibid., 4.

20. Ibid., 5.

21. Robert Block and Christopher Bellamy, "Croats Destroy Mostar's Historic Bridge," *Independent*, November 9, 1993, accessed August 2, 2016, http://www.independent.co.uk/news/croats-destroy-mostars -historic-bridge-1503338.html.

22. The term "War on Terror" was instated by U.S. president George W. Bush after 9/11. In 2013 President Barack Obama publicly declared that the United States was no longer pursuing a War on Terror. See The White House, Office of the Press Secretary, "Remarks by the President at the National Defense University," May 23, 2013, accessed August 2, 2016, https://www.whitehouse.gov/the-press-office/2013/05 /23/remarks-president-national-defense-university.

23. Oleg Orlov, "Ukraine's Forgotten City Destroyed by War," *Guardian*, January 7, 2015, accessed August 2, 2016, https://www.theguardian .com/world/2015/jan/07/-sp-ukraine-pervomaisk-luhansk-forgotten -city-destroyed-by-war.

24. United Nations Radio, "More Than 1.5 Billion People Still Live in Conflict-Affected Countries," February 27, 2013, accessed August 8, 2016, http://www.unmultimedia.org/radio/english/2013/02/more -than-1-5-billion-people-still-live-in-conflict-affected-countries -escap/.

25. UNHCR, *Global Trends: Forced Displacement in 2015*, accessed August 9, 2016, https://s3.amazonaws.com/unhcrsharedmedia/2016/2016-06 -20-global-trends/2016-06-14-Global-Trends-2015.pdf.

26. Ibid.

27. Withnall, "Global Peace Index 2016," 3.

28. Ibid., 24.

29. Stephen Graham, *Cities Under Siege: The New Military Urbanism* (London: Verso, 2010), xxiv.

30. Amnesty International, "Syrian and Russian Forces Targeting Hospitals as a Strategy of War," press release, March 3, 2016, accessed August 9, 2016, https://www.amnesty.org/en/press-releases/2016/03/syrian-and-russian-forces-targeting-hospitals-as-a-strategy-of-war/.

31. "Kosovo Cuts Pristina Water Supply over Alleged ISIS Plot to Poison Reservoir," *Guardian*, July 11, 2015, accessed August 9, 2016, https://www.theguardian.com/world/2015/jul/11/kosovo-cuts-pristina-water-supply-over-alleged-isis-plot-to-poison-reservoir.

32. United Nations Conference on Trade and Development, "Report on UNCTAD Assistance to the Palestinian People: Developments in the Economy of the Occupied Palestinian Territory" (July 6, 2015), 9, accessed August 9, 2016, http://unctad.org/en/PublicationsLibrary/tdb62d3_en.pdf.

33. Ibid., 12.

34. Ibid., 11.

35. Ibid., 9.

36. Ibid., 11.

37. The National Geographic Society reports, "Rain forests that once grew over 14 percent of the land on Earth now cover only about 6 percent and if current deforestation rates continue, these critical habitats could disappear from the planet completely within the next hundred years" (National Geographic Society, "Rain Forest Threats," accessed August 11, 2016, http://environment.nationalgeographic.com/environment/habitats/rainforest-threats/).

38. United Nations Educational, Scientific and Cultural Organization, "Director-General of UNESCO Irina Bokova Firmly Condemns the Destruction of Palmyra's Ancient Temple of Baashamin, Syria," *UNESCO News*, August 24, 2015. Accessed August 11, 2016, http://whc.unesco.org/en/news/1339/.

39. For more on the history of Nimrud and Palmyra, see Charles Gates, *Ancient Cities: The Archaeology of Urban Life in the Ancient Near East and Egypt, Greece, and Rome*, 2nd ed. (Oxon: Routledge, 2011), 170–71, 399–406.

40. United Nations Educational, Scientific and Cultural Organization, *Site of Palmyra*, accessed August 11, 2016, http://whc.unesco.org/en/list/23.

41. Alain Badiou and Peter Engelmann, *Philosophy and the Idea of Communism*, trans. Susan Spitzer (Cambridge: Polity Press, 2015), 19.

42. Ibid., 16–17.

43. There were numerous reports throughout the international media regarding the pillaging and destruction of ancient sites at the hands of ISIS. A quick summary was provided in Andrew Curry, "Here Are the Ancient Sites ISIS Has Damaged or Destroyed," *National Geographic*, September 1, 2015, accessed August 11, 2016, http://news .nationalgeographic.com/2015/09/150901-isis-destruction-looting -ancient-sites-iraq-syria-archaeology/.

44. Rossi, *Architecture of the City*, 127.

45. Badiou and Engelmann, *Philosophy*, 25.

46. Irina Bokova, "Syria: The Director-General Condemns New Destruction at the World Heritage Site of the Old City of Aleppo," *UNESCO News*, accessed August 9, 2016, http://en.unesco.org/news/syria-director -general-condemns-new-destruction-world-heritage-site-old-city -aleppo-o?language=en.

47. Rossi, *Architecture of the City*, 128.

48. Ibid.

49. Ibid.

50. Badiou and Engelmann, *Philosophy*, 42.

51. Henri Bergson, *Creative Evolution*, trans. Arthur Mitchell (New York: Dover, 1998).

52. B'Tselem and Eyal Weizman, "Map of Israeli Settlements in the West Bank," in *A Civilian Occupation*, ed. Rafi Segal and Eyal Weizman (London: Verso, 2003), 109.

53. Jane Bennett, *Vibrant Matter: A Political Ecology of Things* (Durham, N.C.: Duke University Press, 2010).

54. Hollis, *Cities*, 6.

55. Graham, *Cities Under Siege*, xxix.

56. UNHCR, *Global Trends*, 5.

8. PROTEST WITHOUT PEOPLE

1. "Paris Terror Attacks: Newspaper Front Pages from Around the World, in Pictures," accessed June 23, 2016, http://www.telegraph.co .uk/news/worldnews/europe/france/11995738/Paris-Terror-Attacks -Newspaper-front-pages-from-around-the-world-in-pictures.html ?frame=3500717.

2. Jacques Rancière, *The Politics of Aesthetics*, trans. Gabriel Rockhill (London: Continuum, 2004), 42.

3. Judith Butler, *Bodies That Matter: On the Discursive Limits of "Sex"* (London: Routledge, 1993).

4. Walter Benjamin, "Imagination," in *Walter Benjamin: Selected Writings, Volume 1, 1913–1926*, ed. Marcus Bullock and Michael W. Jennings (Cambridge, Mass.: Belknap Press, 1996), 281.

5. Hannah Arendt, *The Human Condition*, 2nd ed. (Chicago: University of Chicago Press, 1998).

6. Fredric Jameson, "Future City," *New Left Review* 21 (May–June 2003), accessed June 1, 2015, http://newleftreview.org/II/21/fredric-jameson-future-city; Slavoj Žižek, *Living in the End Times* (London: Verso, 2011), 334.

7. Brad Evans and Henry Giroux, *Disposable Futures: The Seduction of Violence in the Age of Spectacle* (San Francisco: City Lights Books, 2015), 15.

8. Butler, *Bodies That Matter*, 140.

9. Ibid.

10. Peter Dauvergne, *Environmentalism of the Rich* (Cambridge, Mass.: MIT Press, 2016).

11. Box Office Mojo, "Fire/Firefighter," accessed June 30, 2016, http://www.boxofficemojo.com/genres/chart/?id=fire.htm.

12. Tamar Liebes, "Television Disaster Marathons: A Danger for Democratic Processes," in *Media, Ritual and Identity*, ed. Tamar Liebes and James Curran, 71–84 (London: Routledge, 1998).

13. Robert Putnam, *Bowling Alone: The Collapse and Revival of American Community* (New York: Simon and Schuster, 2000), 243.

14. Jacques Rancière, *Figures of History*, trans. Julie Rose (Cambridge: Polity Press, 2014), 7.

15. Slavoj Žižek, *Violence: Six Sideways Reflections* (New York: Picador, 2008), 2.

16. Guy Debord, *The Society of the Spectacle*, trans. Donald Nicholson-Smith (New York: Zone Books, 1994), paragraph 21.

17. Lydia Saad and Jeffrey M. Jones, "U.S. Concern About Global Warming at Eight-Year High," *Gallup: Politics*, March 16, 2016, accessed June 30, 2016, http://www.gallup.com/poll/190010/concern

-global-warming-eight-year-high.aspx?g_source=CATEGORY
_CLIMATE_CHANGE&g_medium=topic&g_campaign=tiles.

18. Stuart Capstick et al., "Public Perceptions of Climate Change in Britain Following the Winter 2013 / 2014 Flooding," Understanding Risk
Research Group, Cardiff University, May 1, 2015, accessed June 30, 2016,
http://psych.cf.ac.uk/understandingrisk/reports/URG%2015-01%20
WinterFlooding.pdf.

19. Richard Wike, "What the World Thinks About Climate Change
in 7 Charts," Pew Research Center, April 18, 2016, accessed July 1, 2016,
http://www.pewresearch.org/fact-tank/2016/04/18/what-the-world
-thinks-about-climate-change-in-7-charts/.

20. Gilles Deleuze, Difference and Repetition, trans. Paul Patton (New
York: Columbia University Press, 1994), 35–39.

21. Ibid., 136.

22. Nathaneal Arnold, "5 Stormy Movies That Reigned at the Box
Office," Movies Cheat Sheet, July 21, 2014 accessed June 30, 2016,
http://www.cheatsheet.com/entertainment/5-stormy-movies-that
-reigned-at-the-box-office.html/?a=viewall.

23. Fox News, "ISIS Steps Up Propaganda War with Plans to Take Baghdad," June 9, 2015, accessed June 1, 2016, http://video.foxnews.com/v
/4285625535001/isis-steps-up-propaganda-war-with-plans-to-take
-baghdad/?#sp=show-clips; Australian Broadcasting Commission,
Nightline, May 7, 2015, accessed June 1, 2016, http://abcnews.go.com/
International/isis-propaganda-machine-sophisticated-prolific-us
-officials/story?id=30888982.

24. Slavoj Žižek, The Sublime Object of Ideology (London: Verso,
1989), 28.

25. Benjamin, "Imagination," 280.

26. Ibid., 281.

27. Jacques Lacan, The Seminar of Jacques Lacan: Book III, The Psychoses,
1955–1956, ed. Jacques-Alain Miller, trans. Russell Grigg (New York:
Norton, 1997), 175.

28. Naomi Klein, The Shock Doctrine: The Rise of Disaster Capitalism
(New York: Picador, 2007).

29. As U.S. Republican presidential candidate Donald Trump stated in
2016, "I think our biggest form of climate change we should worry

about is nuclear weapons . . . The biggest risk to the world, to me—I know President Obama thought it was climate change—to me the biggest risk is nuclear weapons. That is climate change. That is a disaster, and we don't even know where the weapons are right now. We don't know who has them. We don't know who's trying to get them. The biggest risk for this world and this country is nuclear weapons, the power of nuclear weapons" (Dennis Brady, "Trump: I'm Not a Big Believer in Man-Made Climate Change," *Washington Post*, March 22, 2016, accessed June 13, 2016, https://www.washingtonpost.com/news/energy -environment/wp/2016/03/22/this-is-the-only-type-of-climate-change -donald-trump-believes-in/).

30. Hannah Arendt, *Between Past and Future*, introduction by Jerome Kohn (London: Penguin, 2002), 202.

31. Jacques Rancière, *The Emancipated Spectator*, trans. Gregory Elliott (London: Verso, 2009), 22.

32. Wilhelm Reich, *The Mass Psychology of Fascism*, ed. Mary Higgins and Chester M. Raphael, 3rd ed. (New York: Farrar, Straus and Giroux, 1970).

33. Žižek, *Violence*, 2.

34. Cornelius Castoriadis, "The Retreat from Autonomy: Post-Modernism as Generalized Conformism," *Thesis Eleven* 31, no. 1 (1992): 14–23.

35. Evans and Giroux, *Disposable Futures*, 27.

36. Chiara Bottici and Benoît Challand, eds., *The Politics of Imagination* (New York: Birbeck Law Press, 2011), 33.

37. Ibid.

38. Deleuze, *Difference and Repetition*.

39. Wendy Brown, *Undoing the Demos: Neoliberalism's Stealth Revolution* (New York: Zone Books, 2015), 201.

40. In the United Kingdom, where television airtime for political campaigning is strictly regulated, the Brexit campaign resorted to scaremongering in an attempt to capture more airtime, "publishing a list of murders and rapes committed by 50 EU criminals in Britain" (Jonathan Bacon, "How the Brexit Campaign Matches Up," *Marketing Week*, May 11, 2016, accessed July 10, 2016, https://www.marketingweek.com /2016/05/11/how-the-brexit-campaigns-match-up/). As of March 2015, Donald Trump's presidential campaign had earned nearly $2 billion

worth of media attention, according to a *New York Times* report; Nicholas Confessore and Karen Yourish, "$2 Billion Worth of Free Media for Donald Trump," *New York Times*, March 15, 2016, accessed July 10, 2016, http://www.nytimes.com/2016/03/16/upshot/measuring-donald-trumps-mammoth-advantage-in-free-media.html?_r=0.

41. Bottici and Challand, *The Politics of Imagination*, 21.

42. Ibid., 36.

43. Immanuel Kant, *Critique of Judgment*, trans. Werner S. Pluhar (Indianapolis: Hackett, 1987), 91. Linda Zerilli also highlights Kant's connection between freedom and imagination; see Linda M. G. Zerilli, "'We Feel Our Freedom': Imagination and Judgment in the Thought of Hannah Arendt," *Political Theory* 33, no. 2 (April 2005): 173. Like the earlier Kant, who viewed the imagination as productive and as the condition of memory and understanding, Deleuze maintained that memory and understanding are different from the imagination. Nudging Kant's view of imagination as productive, Deleuze argued that the imagination is a passive synthesis of time, an unconscious contraction that constitutes the live time of the present. Memory and understanding, on the other hand, are active syntheses of time. He further argued that the imagination is imbricated in processes of actualization. Deleuze wrote: "While it is thought which must explore the virtual down to the ground of its repetitions, it is imagination which must grasp the process of actualization from the point of view of these echoes or reprises. It is imagination which crosses domains, orders and levels, knocking down the partitions coextensive with the world, guiding our bodies and inspiring our souls, grasping the unity of mind and nature; a larval consciousness which moves endlessly from science to dream and back again" (Deleuze, *Difference and Repetition*, 220).

44. For Kant, understanding unifies the pure synthesis of imagination out of which knowledge occurs. All in all, Kant understood the imagination to be an individual faculty that synthesizes different intuitions in a unified image. Kant writes in the first edition of the *Critique of Pure Reason*, "[There] are however three sources (capacities or faculties of the soul), which contain the condition of the possibility of all experience, and cannot themselves be derived from any other faculty

of the mind, namely sense, imagination, and apperception. On these are grounded 1) the synopsis of the manifold *a priori* through sense; 2) the synthesis of this manifold through the imagination; finally 3) the unity of this synthesis through original apperception. In addition to their empirical use, all of these faculties have a transcendental one, which is concerned solely with form, and which is possible *a priori*" (B127 and A95). See Immanuel Kant, *Critique of Pure Reason*, trans. and ed. Paul Guyer and Allen W. Wood, 1st ed. (Cambridge: Cambridge University Press, 1998), 225. Kant explains in the same work, "Synthesis in general is, as we shall subsequently see, the mere effect of the imagination, of a blind though indispensable function of the soul, without which we would have no cognition at all, but of which we are seldom even conscious. Yet to bring this synthesis to concepts is a function that pertains to the understanding, and by means of which it first provides cognition in the proper sense" (A78). He continues, "Now pure synthesis, generally represented, yields the pure concept of the understanding" (B104); ibid., 211. Bottici has pointed out that, by the second edition of this work (1787), Kant had modified this productive view of imagination, consigning it to an "intermediary role between intellect and intuition" (Bottici and Challand, *The Politics of Imagination*, 20).

45. For a thorough analysis of the later teachings of Lacan, see Véronique Voruz and Bogdan Wolf, eds., *The Later Lacan: An Introduction* (Albany: SUNY Press, 2007).

46. Bottici and Challand, *The Politics of Imagination*, 29.

47. Zerilli, "'We Feel Our Freedom,'" 174.

48. Deleuze, *Difference and Repetition*, 136.

49. Arendt, *The Human Condition*, 246.

50. Hannah Arendt, *The Promise of Politics*, ed. Jerome Kohn (New York: Schocken Books, 2005).

51. Arendt, *The Human Condition*, 245.

52. Ibid., 237.

53. Ibid.

54. Ibid.

55. Rancière, *The Politics of Aesthetics*, 42. I am also thinking of Kant's view that aesthetic judgments are not simply subjective, which is not tantamount to claiming that a person can't provide reasons why, for example,

they find a particular picture beautiful or not, only that the reasons they provide can't prove the picture is beautiful. For Kant a judgment of beauty, for instance, invokes a shared sensibility and that others concur with that judgment. It would be nonsensical to claim that a picture is beautiful just for me (CJ, 7, p. 55), but this doesn't mean that the picture is conclusively beautiful—not everyone might agree.

56. Rancière, *The Politics of Aesthetics*.

57. Ibid., 43.

58. Ibid.

59. Although Husserlian phenomenology influenced Arendt, her position on intersubjectivity was distinct from Husserl's. He described intersubjectivity through the lens of a transcendental ego. He argued that intersubjectivity enables us to appreciate how other people are like us. Arendt also departed from Husserl's view of a common world as that which people collectively perceive or understand as common.

60. Benedict de Spinoza, *Ethics*, ed. and trans. Edwin Curley (London: Penguin, 1996), 81.

61. Rosi Braidotti, *The Posthuman* (Cambridge: Polity Press, 2013).

62. Michael Hardt and Antonio Negri, *Commonwealth* (Cambridge, Mass.: Belknap Press, 2009), 133.

63. See Brad Evans and Julian Reid, *Resilient Life: The Art of Living Dangerously* (Cambridge: Polity Press, 2014).

64. Paul Virilio, Diller Scofidio + Renfro, Mark Hansen, Laura Kurgan, and Ben Rubin, with Robert Gerard Pietrusko, and Stewart Smith, *EXIT* (2008–2015), 0:44, https://www.youtube.com/watch?v=kyMbF2uuSIw.

65. Artists 4 Climate, Paris 2015, accessed June 1, 2016, http://www.artists4climate.com/en/artists/.

66. Castoriadis, "The Retreat from Autonomy," 14–23.

9. SO TO SPEAK

1. Michael Hardt and Antonio Negri, *Commonwealth* (Cambridge, Mass.: Belknap Press, 2009), 152–53.

2. David Harvey, *Rebel Cities: From the Right to the City to Urban Revolution* (London: Verso, 2012).

3. Jodi Dean, *Crowds and Party* (London: Verso, 2016).

4. Between 2007 and 2009 only 15 percent of funding benefited marginalized communities, the communities most impacted by environmental harms; Sarah Hansen, *Cultivating the Grassroots*, National Committee for Responsible Funders, February 2012, accessed March 1, 2014, https://www.ncrp.org/wp-content/uploads/2016/11/Cultivating_the_grassroots_final_lowres.pdf.

5. Michael Hardt and Antonio Negri, *Empire* (Cambridge, Mass.: Harvard University Press, 2000), 232.

6. Hardt and Negri, *Commonwealth*, viii.

7. Michael Hardt and Antonio Negri, *Multitude: War and Democracy in the Age of Empire* (New York: Penguin, 2004), 103.

8. Ibid., 103, 348.

9. Ibid., 348.

10. Slavoj Žižek, *The Parallax View* (Cambridge, Mass.: MIT Press, 2006), 349.

11. Simon Critchley, "The Problem of Hegemony," cited in Žižek, *Parallax View*, 332.

12. Country Land and Business Association. "UK Votes for Brexit: CLA Analysis of Rural Voting," accessed March 18, 2017, https://www.cla.org.uk/latest/lobbying/brexit-new-opportunities/brexit-news/eu-referendum-cla-analysis-rural-voting.

13. Shearer, Richard. "The Small Town–Big City Split That Elected Donald Trump," Brookings Institution, November 11, 2016, accessed March 18, 2017. https://www.brookings.edu/blog/the-avenue/2016/11/11/the-small-town-big-city-split-that-elected-donald-trump/.

14. Hardt and Negri, *Multitude*, 354.

15. Žižek, *The Parallax View*, 264.

16. Dean, *Crowds and Party*, 253. Nancy Fraser, in "Feminism, Capitalism, and the Cunning of History," *New Left Review* 56 (March/April 2009): 130–47," mounts a cogent and critical assessment of the failure of feminist identity politics under neoliberal capitalism.

17. Ibid., 253.

18. Ibid., 253–54.

19. Dean, *Crowds and Party*, 257.

20. Žižek, *The Parallax View*, 38.

21. Gilles Deleuze and Félix Guattari, *Anti-Oedipus: Capitalism and Schizophrenia*, trans. Robert Hurley, Mark Seem, and Helen R. Lane (Minneapolis: University of Minnesota Press, 1996), 180.

22. Ibid., 18.

23. Ibid., 17.

24. Erich Fromm, *Escape from Freedom* (New York: Holt, 1969), 3.

25. Slavoj Žižek, *Welcome to the Desert of the Real* (London: Verso, 2002), 152.

26. Žižek, *The Parallax View*, 383.

27. Ibid.

28. Thomas Fisher, preface to *New Directions in Sustainable Design*, ed. Adrian Parr and Michael Zaretsky (Oxon: Routledge, 2011), xvii.

29. Slavoj Žižek, *Looking Awry: An Introduction to Jacques Lacan Through Popular Culture* (Cambridge, Mass.: MIT Press, 1992), 36.

30. Žižek, *Desert of the Real*, 152.

31. Ibid., 153.

32. Arundhati Roy, *The Cost of Living* (Toronto: Modern Library, 1999), 25.

33. Žižek, *Looking Awry*, 77.

34. Oxfam. "62 People Own Same as Half World," *Press Releases*, January 18, 2016, accessed July 1, 2016, http://www.oxfam.org.uk/media-centre /press-releases/2016/01/62-people-own-same-as-half-world-says -oxfam-inequality-report-davos-world-economic-forum.

BIBLIOGRAPHY

Agamben, Giorgio. *Homo Sacer: Sovereign Power and Bare Life.* Trans. Daniel Heller-Roazen. Stanford, Calif.: Stanford University Press, 1998.

Amnesty International. *Kenya: The Unseen Majority; Nairobi's Two Million Slum Dwellers.* 2009. Accessed February 3, 2016. https://www.amnesty.nl/sites/default/files/public/rap_kenia_the_unseen_majority.pdf.

——. "Syrian and Russian Forces Targeting Hospitals as a Strategy of War." Press release, March 3, 2016. Accessed August 9, 2016. https://www.amnesty.org/en/press-releases/2016/03/syrian-and-russian-forces-targeting-hospitals-as-a-strategy-of-war/.

Angel, Shlomo. *Planet of Cities.* Cambridge, Mass.: Lincoln Institute of Land Policy, 2012.

"Annan Lays Out Detailed Five-Point UN Strategy to Combat Terrorism." *UN News Centre*, March 10, 2005. Accessed December 22, 2015. http://www.un.org/apps/news/story.asp?NewsID=13599&#.VKGu88AAvA.

Aquinas, Thomas. *Summa Theologiae.* Vol. 38, *Injustice.* Ed. Marcus Lefébure. Cambridge: Cambridge University Press, 2006.

Arendt, Hannah. *Between Past and Future.* Introduction by Jerome Kohn. London: Penguin, 2002.

——. *The Human Condition.* 2nd ed. Chicago: University of Chicago Press, 1998.

——. *The Promise of Politics.* Ed. Jerome Kohn. New York: Schocken Books, 2005.

Arluke, Arnold, and Clinton Sanders. *Regarding Animals.* Philadelphia: Temple University Press, 2012.

Arnold, Nathaneal. "5 Stormy Movies That Reigned at the Box Office." *Movies Cheat Sheet*, July 21, 2014. Accessed June 30, 2016. http://www .cheatsheet.com/entertainment/5-stormy-movies-that-reigned-at-the -box-office.html/?a=viewall.

Artists 4 Climate. Paris 2015. Accessed June 1, 2016. http://www .artists4climate.com/en/artists/.

Australian Broadcasting Commission. *Nightline*, May 7, 2015. Accessed June 1, 2016. http://abcnews.go.com/International/isis-propaganda -machine-sophisticated-prolific-us-officials/story?id=30888982.

Bacon, Jonathan. "How the Brexit Campaign Matches Up." *Marketing Week*, May 11, 2016. Accessed July 10, 2016. https://www.marketingweek .com/2016/05/11/how-the-brexit-campaigns-match-up/.

Badiou, Alain. *Being and Event*. Trans. Oliver Feltham. London: Continuum, 2007.

——. *Ethics: An Essay on the Understanding of Evil*. Trans. Peter Hallward. London: Verso 2001.

Badiou, Alain, and Peter Engelmann. *Philosophy and the Idea of Communism*. Trans. Susan Spitzer. Cambridge: Polity Press, 2015.

Balanyá, Belén, and Oscar Reyes. "Caught in the Cross-Hairs: How Industry Lobbyists Are Gunning for EU Climate Targets." Corporate Europe Observatory and Carbon Trade Watch, June 2011. Accessed June 10, 2014. http://www.carbontradewatch.org/downloads/publications/Caught_ in_the_cross_hairs.pdf.

Bandt, Adam. "Had We but World Enough and Time." *Australian Feminist Law Journal* 31, no. 1 (2009): 15–32.

Beirich, Heidi. *Greenwash: Nativists, Environmentalism, and the Hypocrisy of Hate*. Ed. Mark Potok. Southern Poverty Law Center, July 2010. Accessed July 21, 2016. https://www.splcenter.org/sites/default/files/d6_ legacy_files/downloads/publication/Greenwash.pdf.

Bell, Bryan, and Katie Wakeford, eds. *Expanding Architecture: Design as Activism*. New York: Metropolis Books, 2008.

Benjamin, Walter. *Walter Benjamin: Selected Writings, Volume 1, 1913–1926*. Ed. Marcus Bullock and Michael W. Jennings. Cambridge, Mass.: Belknap Press, 1996.

——. "The Work of Art in the Age of Mechanical Reproduction." In *Illuminations*, trans. Harry Zohn, 219–53. London: Cape, 1970.

Bennett, Jane. *Vibrant Matter: A Political Ecology of Things*. Durham, N.C.: Duke University Press, 2010.

Bergson, Henri. *Creative Evolution*. Trans. Arthur Mitchell. New York: Dover, 1998.

Berube, Alan, Audrey Singer, and William Frey. *The State of Metropolitan America*. Brookings Institution, 2010. Accessed March 21, 2011. http://www.brookings.edu/~/media/research/files/reports/2010/5/09%20metro%20america/metro_america_report.pdf.

Biehl, Janet, and Peter Staudenmaier. *Ecofascism: Lessons from the German Experience*. Oakland, Calif.: AK Press, 1995.

Black Sun Invictus. "Interview with Troy Southgate." *National-Anarchist Movement*, May 29, 2012. Accessed July 20, 2016. http://www.national-anarchist.net/2012/05/interview-with-troy-southgate-from.html.

Block, Robert, and Christopher Bellamy. "Croats Destroy Mostar's Historic Bridge." *Independent*, November 9, 1993. Accessed August 2, 2016. http://www.independent.co.uk/news/croats-destroy-mostars-historic-bridge-1503338.html.

Bloomberg, Michael, Henry Paulson, and Thomas Steyer. *Risky Business: The Economic Risks of Climate Change in the United States*, 2014. Accessed June 28, 2014. http://riskybusiness.org/uploads/files/RiskyBusiness_PrintedReport_FINAL_WEB_OPTIMIZED.pdf.

Blowers, Andrew. "Transition or Transformation? Environmental Policy Under Thatcher." *Public Administration* 65, no. 3 (1987): 277–94.

Bokova, Irina. "Syria: The Director-General Condemns New Destruction at the World Heritage Site of the Old City of Aleppo." *UNESCO News*. Accessed August 9, 2016. http://en.unesco.org/news/syria-director-general-condemns-new-destruction-world-heritage-site-old-city-aleppo-o?language=en.

Bookchin, Murray. *The Ecology of Freedom: The Emergence and Dissolution of Hierarchy*. Palo Alto, Calif.: Cheshire Books, 1982.

——. *The Next Revolution: Popular Assemblies and the Promise of Direct Democracy*. Ed. Debbie Bookchin and Blair Taylor. London: Verso, 2015.

——. *The Philosophy of Social Ecology: Essays on Dialectical Naturalism*. 2nd ed. Montreal: Black Rose Books, 1995.

——. *Remaking Society: Pathways to a Green Future*. Boston: South End Press, 1990.

——. *Social Ecology and Communalism.* Oakland, Calif.: AK Press, 2006.

Bottici, Chiara, and Benoît Challand, eds. *The Politics of Imagination.* New York: Birbeck Law Press, 2011.

Box Office Mojo. "Fire/Firefighter." Accessed June 30, 2016. http://www.boxofficemojo.com/genres/chart/?id=fire.htm.

Braidotti, Rosi. *The Posthuman.* Cambridge: Polity Press, 2013.

Bramwell, Anna. *Blood and Soil: Richard Walther Darré and Hitler's "Green Party."* Bourne End, Buckinghamshire, U.K.: Kensal Press, 1984.

Brenner, Neil. "Theses on Urbanization." *Public Culture* 25, no. 1 (2013): 85–114.

Bretton Woods Project. "Analysis of World Bank Voting Reforms." Briefing, April 30, 2010. Accessed February 1, 2016. http://www.brettonwoodsproject.org/2010/04/art-566281/.

British Broadcasting Corporation. "Europe in Housing Market 'Agony,' Says Rics." *BBC Business News*, February 28, 2012. Accessed December 21, 2012. http://www.bbc.co.uk/news/business-17191691.

British Petroleum. *Alternative Energy*, 2014. Accessed February 6, 2014. http://www.bp.com/en/global/corporate/sustainability/the-energy-future/alternative-energy.html.

Brown, Paul. "Protest by CND Stretches 14 Miles." *Guardian*, April 2, 1983. Accessed July 20, 2016. https://www.theguardian.com/fromthearchive/story/0,,1866956,00.html.

Brown, Wendy. *Undoing the Demos: Neoliberalism's Stealth Revolution.* New York: Zone Books, 2015.

Brune, Michael. "Sierra Club Statement on Police Killings of Alton Sterling and Philando Castile." *Sierra Club*, July 7, 2016. Accessed July 17, 2016. http://content.sierraclub.org/press-releases/2016/07/sierra-club-statement-police-killings-alton-sterling-and-philando-castile.

B'Tselem and Eyal Weizman. "Map of Israeli Settlements in the West Bank." In *A Civilian Occupation*, ed. Rafi Segal and Eyal Weizman, 108–19. London: Verso, 2003.

Buffington, Larry. "Professionalization: A Strategy for Improving Environmental Management." *Humboldt Journal of Social Relations* 2, no. 1 (fall/winter 1974): 18–21.

Bullard, Robert, ed. *Just Sustainabilities: Development in an Unequal World.* Cambridge, Mass.: MIT Press, 2003.

——. *The Quest for Environmental Justice: Human Rights and the Politics of Pollution.* San Francisco: Sierra Club Books, 2005.

Burns, Carol, and Andrea Kahn, eds. *Site Matters: Designs, Concepts, Histories, and Strategies.* New York: Routledge, 2005.

Butler, Judith. *Bodies That Matter: On the Discursive Limits of "Sex."* London: Routledge, 1993.

Butler, Judith, and Athena Athanasiou. *Dispossession: The Performative in the Political.* Cambridge: Polity Press, 2013.

Capstick, Stuart B., et al. "Public Perceptions of Climate Change in Britain Following the Winter 2013/2014 Flooding." Understanding Risk Research Group, Cardiff University, May 1, 2015. Accessed June 30, 2016. http://psych.cf.ac.uk/understandingrisk/reports/URG%2015-01%20 WinterFlooding.pdf.

Castoriadis, Cornelius. "The Retreat from Autonomy: Post-Modernism as Generalized Conformism." *Thesis Eleven* 31, no. 1 (1992): 14–23.

Center for Public Integrity and International Consortium of Investigative Journalists. *Promoting Privatization*, 2003. Accessed May 5, 2014. http:projects.publicintegrity.org/water/report.aspx?aid=45.

"Changes in Atmospheric Carbon Dioxide, Methane, and Nitrous Oxide." In *IPCC Fourth Assessment Report: Climate Change 2007.* Accessed March 17, 2017. https://www.ipcc.ch/publications_and_data/ar4/wg1/ en/tssts-2-1-1.html.

Chivian, Eric, and Aaron Bernstein, eds. *Sustaining Life: How Human Health Depends on Biodiversity.* New York: Center for Health and the Global Environment/Oxford University Press, 2008.

Christie, Les. "Foreclosures up a Record 81% in 2008." *CNN Money*, January 15, 2009. Accessed December 22, 2012. http://money.cnn.com/2009/01/15/ real_estate/millions_in_foreclosure/.

Clark, Lara P., Dylan B. Millet, and Julian D. Marshall. "National Patterns in Environmental Injustice and Inequality: Outdoor NO_2 Air Pollution in the United States." *PLOS ONE* 9, no. 4 (April 2014): 1–8.

Cline, Seth. "Sheldon Adelson Spent \$150 Million on Election." *US News*, December 3, 2012. Accessed June 11, 2014. http://www.usnews.com/ news/articles/2012/12/03/sheldon-adelson-ended-up-spending-150 -million.

Cole, Steve, and Leslie McCarthy. "Long-Term Warming Likely to Be Significant Despite Recent Slowdown." *NASA*, March 11, 2014. Accessed March 22, 2014. http://climate.nasa.gov/news/1050.

Coleman, James. "Social Capital in the Creation of Human Capital." *American Journal of Sociology* 94, no. 51 (1988): 95–120.

Confessore, Nicholas, and Karen Yourish. "$2 Billion Worth of Free Media for Donald Trump." *New York Times*, March 15, 2016. Accessed July 10, 2016. http://www.nytimes.com/2016/03/16/upshot/measuring-donald -trumps-mammoth-advantage-in-free-media.html?_r=0.

Co-operative Housing International. "About Kenya." Accessed February 14, 2016. http://www.housinginternational.coop/co-ops/kenya.

Copeland, Brian, and M. Scott Taylor. "Trade, Growth, and the Environment." *Journal of Economic Literature* 42, no. 1 (2004): 7–71.

Corburn, Jason, and Chantal Hildebrand. "Slum Sanitation and the Social Determinant of Women's Health in Nairobi, Kenya." *Journal of Environmental and Public Health* 2015 (2015): 1–6.

Corporate Eco Forum. *The New Business Imperative: Valuing Natural Capital*, 2012. Accessed January 4, 2014. http://www.corporateecoforum.com /valuingnaturalcapital/offline/download.pdf.

Corporate Europe Observatory. "The EU: A Hollow Champion for the Climate," December 7, 2009. Accessed June 10, 2014. http://corporateeu rope.org/sites/default/files/sites/default/files/files/article/eu_hollow_ champion_dec_09.pdf.

Country Land and Business Association. "UK Votes for Brexit: CLA Analysis of Rural Voting." Accessed March 18, 2017. https://www.cla .org.uk/latest/lobbying/brexit-new-opportunities/brexit-news/eu -referendum-cla-analysis-rural-voting.

Cruz, Wilfrido, and Robert Repetto. *The Environmental Effects of Stabilization and Structural Adjustment Programs: The Philippines Case.* Washington, D.C.: World Resources Institute, 1992.

Curry, Andrew. "Here Are the Ancient Sites ISIS Has Damaged or Destroyed." *National Geographic*, September 1, 2015. Accessed August 11, 2016. http://news.nationalgeographic.com/2015/09/150901-isis-destruction -looting-ancient-sites-iraq-syria-archaeology/.

Dale, Gareth, Manu Mathai, and Jose Puppim de Oliveira, eds. *Green Growth: Ideology, Political Economy, and the Alternatives.* London: Zed Books, 2016.

Dauvergne, Peter. *Environmentalism of the Rich.* Cambridge, Mass.: MIT Press, 2016.

Davidson, Helen, and Nick Evershed. "Australia Bushfires Live: Fears Blue Mountains Fires Will Join Together." *Guardian*, October 21, 2013. Accessed December 16, 2014. http://www.theguardian.com/world/2013 /oct/21/nsw-fires-residents-evacuate.

Davis, Mike. *Planet of Slums*. London: Verso, 2006.

Dean, Jodi. *Crowds and Party*. London: Verso, 2016.

Dean, Judith. "Does Trade Liberalization Harm the Environment? A New Test." *Canadian Journal of Economics* 35, no. 4 (2002): 819–42.

Debord, Guy. *The Society of the Spectacle*. Trans. Donald Nicholson-Smith. New York: Zone, 1999.

Deleuze, Gilles. *Difference and Repetition*. Trans. Paul Patton. New York: Columbia University Press, 1994.

Deleuze, Gilles, and Félix Guattari. *Anti-Oedipus: Capitalism and Schizophrenia*. Trans. Robert Hurley, Mark Seem, and Helen R. Lane. London: Continuum, 2004.

——. *A Thousand Plateaus: Capitalism and Schizophrenia*. Trans. Brian Massumi. Minneapolis: University of Minnesota Press, 1987.

——. *What Is Philosophy?*. Trans. Hugh Tomlinson and Graham Burchell. New York: Columbia University Press, 1994.

Dempsey, Bobbi, and Todd Beitler. *The Complete Idiot's Guide to Buying Foreclosures*. 2nd ed. New York: Penguin, 2005.

Dennis, Brady. "Trump: I'm Not a Big Believer in Man-Made Climate Change." *Washington Post*, March 22, 2016. Accessed June 13, 2016. https://www.washingtonpost.com/news/energy-environment/wp/2016 /03/22/this-is-the-only-type-of-climate-change-donald-trump-beli eves-in/.

Derrida, Jacques. *Writing and Difference*. Trans. Alan Bass. London: Routledge, 2001.

Dobson, Andrew. *Green Political Thought*. 2nd ed. London: Routledge, 1995.

Dominick, Raymond H. *The Environmental Movement in Germany: Prophets and Pioneers, 1871–1971*. Bloomington: Indiana University Press, 1992.

Dryzek, John. S. *The Politics of the Earth*. 2nd ed. Oxford: Oxford University Press, 2005.

Eckersley, Robyn. *Environmentalism and Political Theory: Toward an Ecocentric Approach*. Albany: SUNY Press, 1992.

Economic and Social Commission for Asia and the Pacific, Asian Development Bank, and United Nations Environment Program. *Green Growth, Resources and Resilience: Environmental Sustainability in Asia and the Pacific*, 2012. Accessed June 6, 2014. https://www.adb.org/sites/default/files/publication/29567/green-growth-resources-resilience.pdf.

Egerton, Frank N. *Roots of Ecology: Antiquity to Haeckel*. Berkeley: University of California Press, 2012.

Ekins, Paul. *Economic Growth, Human Welfare and Environmental Sustainability: The Prospects for Green Growth*. London: Routledge, 2000.

Embassy of the Bolivarian Republic of Venezuela to the U.S. *Fact Sheet: Urban Agriculture in Venezuela*, August 2012. Accessed March 1, 2013. http://venezuela-us.org/live/wp-content/uploads/2009/08/08.20.2012-Urban-Agriculture-ENG1.pdf.

Escobar, Arturo. *Encountering Development: The Making and Unmaking of the Third World*. Princeton, N.J.: Princeton University Press, 2012.

Esteva, Gustavo, Salvatore Babones, and Philipp Babcicky. *The Future of Development: A Radical Manifesto*. Bristol, U.K.: Policy Press, 2013.

EurActiv. "EU-Canada Free Trade Deal 'Opens Door to Environmental Lawsuits,'" 2014. Accessed May 10, 2014. http://www.euractiv.com/trade/eu-canada-free-trade-deal-opens-news-533400.

EUROFER. "Open Letter of the EU Steel Industry to the Governments of the EU Member States, the European Parliament and the European Commission," May 19, 2011. Accessed June 11, 2014. http://www.eurofer.org/News%26Events/Press%20releases/OpenLetter2.fhtml.

Eurostat. "Glossary: Degree of Urbanisation." *Statistics Explained*. Accessed August 8, 2016. http://ec.europa.eu/eurostat/statistics-explained/index.php/Glossary:Degree_of_urbanisation.

Evans, Brad, and Henry Giroux. *Disposable Futures: The Seduction of Violence in the Age of Spectacle*. San Francisco: City Lights Books, 2015.

Evans, Brad, and Julian Reid. *Resilient Life: The Art of Living Dangerously*. Cambridge: Polity Press, 2014.

Evans, Rob, and Paul Lewis. "Revealed: How Energy Firms Spy On Environmental Activists." *Guardian*, February 14, 2011. Accessed December 21, 2014. http://www.theguardian.com/environment/2011/feb/14/energy-firms-activists-intelligence-gathering.

Fairhall, David. *Common Ground: The Story of Greenham*. London: St. Martin's Press, 2000.

Fanon, Frantz. *Black Skin, White Masks*. Trans. Richard Philcox. New York: Grove Press, 2008.

Food and Agriculture Organization of the United Nations. *The Water-Energy-Food Nexus: A New Approach in Support of Food Security and Sustainable Agriculture*, June 2014. Accessed March 18, 2017. http://www.fao .org/nr/water/docs/FAO_nexus_concept.pdf.

Foster, John Bellamy, Brett Clark, and Richard York. *The Ecological Rift: Capitalism's War on the Earth*. New York: Monthly Review Press, 2010.

Foster, Sheila. "The City as an Ecological Space: Social Capital and Urban Land Use." *Notre Dame Law Review* 82, no. 2 (2006): 527–82.

Foucault, Michel. "About the Beginning of the Hermeneutics of the Self." *Political Theory* 21, no. 2 (1993): 198–227.

——. *The Birth of Biopolitics: Lectures at the Collège de France, 1978–1979*. Trans. Graham Burchell. New York: Palgrave Macmillan, 2008.

——. *Discipline and Punish: The Birth of the Prison*. Trans. Alan Sheridan. New York: Vintage, 1995.

——. *Language, Counter-Memory, Practice: Selected Essays and Interviews*. Trans. Donald F. Bouchard and Sherry Simon. Ithaca, N.Y.: Cornell University Press, 1977.

——. "Questions of Method." In *The Foucault Effect: Studies in Governmentality*, ed. Graham Burchell, Colin Gordon, and Peter Miller, 73–86. Chicago: University of Chicago Press, 1991.

——. "The Subject and Power." *Critical Inquiry* 8, no. 4 (summer 1982): 777–95.

Fox News. "ISIS Steps Up Propaganda War with Plans to Take Baghdad," June 9, 2015. Accessed June 1, 2016. http://video.foxnews.com/v/428562 5535001/isis-steps-up-propaganda-war-with-plans-to-take-baghdad/?#sp =show-clips.

Frankfurt School of Finance and Management and UNEP Collaborating Centre for Climate and Sustainable Energy Finance. *Global Trends in Renewable Energy Investment, 2012*. Accessed January 1, 2014. http://fs-unep -centre.org/sites/default/files/publications/globaltrendsreport2012.pdf.

Fraser, Nancy. "Feminism, Capitalism, and the Cunning of History." *New Left Review* 56 (March/April 2009): 130–47.

Frichot, Hélène, and Stephen Loo, eds. *Deleuze and Architecture*. Edinburgh: Edinburgh University Press, 2013.

Fromm, Erich. *Escape from Freedom*. New York: Holt, 1969.

Fyodorov, Miron. "Interview with Troy Southgate for *Kinovar* (Russia)." *Euro-Synergies*, February 5, 2008. Accessed July 17, 2016. http://www .thephora.net/forum/archive/index.php/t-33981.html.

Gasman, Daniel. *The Scientific Origins of National Socialism*. London: Transaction, 2007.

Gates, Charles. *Ancient Cities: The Archaeology of Urban Life in the Ancient Near East and Egypt, Greece, and Rome*. 2nd ed. Oxon: Routledge, 2011.

George, Susan. *A Fate Worse Than Debt: The Financial Crisis and the Poor*. New York: Grove Press, 1988.

Gier, Nicholas F. "Confucius, Gandhi and the Aesthetics of Virtue." *Asian Philosophy* 11, no. 1 (2001): 41–54.

Gilens, Martin, and Benjamin I. Page. "Testing Theories of American Politics: Elites, Interest Groups, and Average Citizens." *Perspectives on Politics* 12, no. 3 (summer 2014): 564–81.

Giroux, Henri. *Disposable Youth: Racialized Memories and the Culture of Cruelty*. London: Routledge, 2012.

Glaeser, Edward. *Triumph of the City: How Urban Spaces Make Us Human*. London: Macmillan, 2011.

Goodin, Robert. *Green Political Theory*. Cambridge: Polity Press, 1992.

Government of Canada. Lone Pine Resources Inc. versus the Government of Canada, September 6, 2013. Accessed June 1, 2014. http://www .international.gc.ca/trade-agreements-accords-commerciaux/assets/ pdfs/disp-diff/lone-02.pdf.

Government of India, Ministry of Home Affairs. "Census of India 2011: Provisional Population Totals; Urban Agglomerations and Cities." Accessed August 8, 2016. http://censusindia.gov.in/2011-prov-results/ paper2/data_files/India2/1.%20Data%20Highlight.pdf.

Graham, Stephen. *Cities Under Siege: The New Military Urbanism*. London: Verso, 2010.

Greenpeace. *Kyoto*. Accessed April 10, 2014. http://www.greenpeace.org/ international/en/campaigns/climate-change/a/governments/kyoto/.

——. *License to Kill: How Deforestation for Palm Oil Is Driving Sumatran Tigers Toward Extinction*, 2013. Accessed January 2, 2015. http://www .greenpeace.org/international/Global/international/publications/ forests/2013/LicenceToKill_ENG_LOWRES.pdf.

———. "U.S. Withdraws from Kyoto Protocol," April 5, 2001. Accessed January 3, 2014. http://www.greenpeace.org/usa/en/news-and-blogs/news/u-s-withdraws-from-kyoto-prot/.

Grice, Andrew. "Energy Rip-Off: 'Big Six' Firms Too Close to Ministers, Says Ed Miliband." *Independent*, October 7, 2013. Accessed June 22, 2014. http://www.independent.co.uk/news/uk/politics/energy-ripoff-big-six-firms-too-close-to-ministers-says-ed-miliband-8862740.html.

Grosz, Elizabeth. *Chaos, Territory, Art*. New York: Columbia University Press, 2008.

Guha-Sapir, Debarati, Philippe Hoyois, and Regina Below. *Annual Disaster Statistical Review 2014: The Numbers and Trends*. Centre for Research on the Epidemiology of Disasters, September 22, 2014. Accessed December 16, 2014. http://www.cred.be/sites/default/files/ADSR_2013.pdf.

Haaga, John, Richard Scott, and Jennifer Hawes-Dawson. *Drug Use in the Detroit Metropolitan Area: Problems, Programs, and Policy Options*. Santa Monica, Calif.: RAND, 1992. Accessed December 20, 2014. http://www.rand.org/content/dam/rand/pubs/reports/2009/R4085.pdf.

Hallegatte, Stéphane, Geoffrey Heal, Marianne Fay, and David Treguer. "From Growth to Green Growth: A Framework." World Bank Group Policy Research Working Papers, WPS5872, November 2011.

Hamilton, Kirsty. *The Oil Industry and Climate Change: A Greenpeace Briefing*. Greenpeace International, 1998. Accessed January 4, 2014. http://www.greenpeace.org/international/Global/international/planet-2/report/2006/3/the-oil-industry-and-climate-c.pdf.

Hansen, James. *Storms of My Grandchildren: The Truth About the Coming Climate Catastrophe and the Chance to Save Humanity*. New York: Bloomsbury, 2011.

Hansen, Sarah. *Cultivating the Grassroots*. National Committee for Responsible Funders, February 2012. Accessed March 1, 2014. https://www.ncrp.org/wp-content/uploads/2016/11/Cultivating_the_grassroots_final_lowres.pdf.

Hardin, Garrett. "The Tragedy of the Commons." *Science* 162, no. 3859 (1968): 1243–48.

Hardt, Michael, and Antonio Negri. *Commonwealth*. Cambridge, Mass.: Belknap Press, 2009.

———. *Empire*. Cambridge, Mass.: Harvard University Press, 2000.

——. *Multitude: War and Democracy in the Age of Empire*. New York: Penguin, 2004.

Hartmann, Betsy. "The Greening of Hate: An Environmentalist's Essay." In *Greenwash: Nativists, Environmentalism, and the Hypocrisy of Hate*, ed. Mark Potok, 13–15. Southern Poverty Law Center, July 2010. Accessed July 21, 2016. https://www.splcenter.org/sites/default/files/d6_legacy_files/downloads/publication/Greenwash.pdf.

Harvey, David. *A Brief History of Neoliberalism*. Oxford: Oxford University Press, 2005.

——. *The Enigma of Capital*. Oxford: Oxford University Press, 2010.

——. *Rebel Cities: From the Right to the City to the Urban Revolution*. London: Verso, 2012.

——. *Spaces of Global Capitalism: A Theory of Uneven Geographical Development*. London: Verso, 2006.

Hawken, Paul, Amory Lovins, and Hunter Lovins. *Natural Capitalism: Creating the Next Industrial Revolution*. Boston: Little, Brown, 1999.

Hayek, Friedrich. *The Road to Serfdom*. Chicago: University of Chicago Press, 1944.

Herzog, Tim. "World Greenhouse Gas Emissions in 2005." World Resources Institute, July 2009. Accessed March 1, 2014. http://www.wri.org/publication/world-greenhouse-gas-emissions-2005.

Heynen, Nik, James McCarthy, Scott Prudham, and Paul Robbins, eds. *Neoliberal Environments: False Promises and Unnatural Consequences*. London: Routledge, 2007.

Hillier, Debbie, and Benedict Dempsey. *A Dangerous Delay: The Cost of Late Response to Early Warnings in the 2011 Drought in the Horn of Africa*. Oxfam International and Save the Children, January 18, 2012. Accessed March 18, 2017. http://policy-practice.oxfam.org.uk/publications/a-dangerous-delay-the-cost-of-late-response-to-early-warnings-in-the-2011-droug-203389.

Hipperson, Sarah. *Greenham Common Women's Peace Camp*. Accessed July 20, 2016. http://www.greenhamwpc.org.uk/.

Hitler, Adolf. *Mein Kampf*. Trans. James Murphy (1924). Accessed July 22, 2016. http://www.greatwar.nl/books/meinkampf/meinkampf.pdf.

Hoffmann, Susanne. *Architecture Is Participation: Die Baupiloten; Methods and Projects*. Berlin: JOVIS, 2015.

Hollis, Leo. *Cities Are Good for You: The Genius of the Metropolis*. New York: Bloomsbury, 2013.

Homer, Sean. *Fredric Jameson: Marxism, Hermeneutics, Post-modernism*. Cambridge: Cambridge University Press, 1998.

Honig, Bonnie. *Emergency Politics: Paradox, Law, Democracy*. Princeton, N.J.: Princeton University Press, 2009.

——. "Three Models of Emergency Politics." *Boundary* 41, no. 2 (2014): 45–70.

"Horn of Africa Food Crisis Remains Dire as Famine Spreads in Somalia." *UN News Centre*, September 5, 2001. Accessed December 20, 2014. http://www.un.org/apps/news/story.asp?NewsID=39450#.VJrs0sAAvA.

Huchzermeyer, Marie. "Slum Upgrading in Nairobi within the Housing and Basic Services Market." *Journal of Asian and African Studies* 43, no. 1 (2008): 19–39.

Human Rights Watch. *The Dark Side of Green Growth*, 2013. Accessed May 22, 2014. http://www.hrw.org/sites/default/files/reports/indonesia0713webw cover_1.pdf.

Ignatieff, Michael. *The Lesser Evil: Political Ethics in an Age of Terror*. Princeton, N.J.: Princeton University Press, 2004.

Index Mundi. "Kenya Demographics Profile 2014." Accessed February 3, 2016. http://www.indexmundi.com/kenya/demographics_profile.html.

Intergovernmental Panel on Climate Change. *Synthesis Report of the Fifth Assessment Report of the Intergovernmental Panel on Climate Change*. Accessed December 20, 2014. https://www.ipcc.ch/pdf/assessment -report/ar5/syr/SYR_AR5_LONGERREPORT.pdf.

International Centre for the Settlement of Investment Disputes. Tecnicas Medioambientales Tecmed S.A. versus the United Mexican States. Case no. ARB (AF)/00/2. May 29, 2003. Accessed April 1, 2014. http://www.italaw.com/documents/Tecnicas_001.

"ISIS Video Appears to Show Beheadings of Egyptian Coptic Christians in Libya." *CNN News*, February 16, 2015. Accessed June 1, 2015. http://www.cnn.com/2015/02/15/middleeast/isis-video-beheadings-christians/.

Jacobs, Jane. *The Death and Life of Great American Cities*. New York: Random House, 1961.

Jameson, Fredric. "Future City." *New Left Review* 21 (May–June 2003). Accessed June 1, 2015. http://newleftreview.org/II/21/fredric-jameson -future-city.

——. "The Politics of Utopia." *New Left Review* 25 (January–February 2004). Accessed May 10, 2015. http://newleftreview.org/II/25/fredric -jameson-the-politics-of-utopia.

——. *Postmodernism; or, The Cultural Logic of Late Capitalism.* Durham, N.C.: Duke University Press, 1992.

Jarboe, James F. "Testimony Before the House Resources Committee, Subcommittee on Forests and Forest Health." *Federal Bureau of Investigation*, February 12, 2002. Accessed December 22, 2014. http://www.fbi .gov/news/testimony/the-threat-of-eco-terrorism.

Jenkins, Craig, and Craig Eckert. "Channeling Black Insurgency: Elite Patronage and Professional Social Movement Organizations in the Development of the Black Movement." *American Sociological Review* 51, no. 6 (December 1986): 812–29.

Johnston, Sadhu Aufochs, Steven S. Nicholas, and Julia Parzen. *The Guide to Greening Cities.* Washington, D.C.: Island Press, 2013.

Kant, Immanuel. *Critique of Judgment.* Trans. Werner S. Pluhar. Indianapolis: Hackett, 1987.

——. *Critique of Pure Reason.* Trans. and ed. Paul Guyer and Allen W. Wood. Cambridge: Cambridge University Press, 1998.

——. *Groundwork of the Metaphysics of Morals.* Ed. Mary Gregor and Jens Timmermann. Rev. ed. Cambridge: Cambridge University Press, 2012.

Kim, Jeong-su. "The Environmental Fallout of the Four Major Rivers Project." *Hankyoreh*, August 3, 2013. Accessed June 8, 2014. http://www.hani .co.kr/arti/english_edition/e_national/598190.html.

Kim, Jim Yong. "Speech by World Bank Group President Jim Yong Kim at the Migration and the Global Development Agenda." *World Bank*, December 9, 2015. Accessed December 26, 2015. http://www.worldbank .org/en/news/speech/2015/12/09/speech-by-world-bank-group-president -jim-yong-kim-at-the-migration-and-the-global-development-agenda.

——. "Statement by World Bank Group President Jim Yong Kim at Spring Meetings 2014 Opening Press Conference." *World Bank*, April 10, 2014. Accessed June 3, 2014. http://www.worldbank.org/en/news/ speech/2014/04/10/statement-world-bank-group-president-jim-yong -kim-spring-meetings-2014-opening-press-conference.

King, Martin Luther, Jr. "The American Dream." Delivered at Ebenezer Baptist Church, Atlanta, Georgia, July 4, 1965. Accessed January 31,

2016. http://kingencyclopedia.stanford.edu/encyclopedia/document sentry/doc_the_american_dream/.

——. "My Pilgrimage to Nonviolence." Martin Luther King, Jr., Papers Project, September 1, 1958. Accessed June 1, 2015. http://kingencyclo pedia.stanford.edu/primarydocuments/Vol4/1-Sept-1958_MyPilgrim ageToNonviolence.pdf.

Klein, Naomi. *The Shock Doctrine: The Rise of Disaster Capitalism*. New York: Picador, 2007.

Konikow, Leonard F., and Eloise Kendy. "Groundwater Depletion: A Global Problem." *Hydrogeology Journal* 13, no. 1 (March 2005): 317–20.

Koont, Sinan. "The Urban Agriculture of Havana." *Monthly Review* 60, no. 8 (2009). Accessed March 1, 2013. http://monthlyreview.org/2009/01/01/ the-urban-agriculture-of-havana/.

"Kosovo Cuts Pristina Water Supply over Alleged ISIS Plot to Poison Reservoir." *Guardian*, July 11, 2015. Accessed August 9, 2016. https://www .theguardian.com/world/2015/jul/11/kosovo-cuts-pristina-water-supply -over-alleged-isis-plot-to-poison-reservoir.

Kristeva, Julia. *Hannah Arendt: Life Is a Narrative*. Trans. Frank Collins. Toronto: University of Toronto Press, 2001.

Krupp, Fred. "Statement of Environmental Defense Fund President Fred Krupp on Recent Events in Baton Rouge, Minneapolis, and Dallas." *Environmental Defense Fund*, July 8, 2016. Accessed July 17, 2016. https:// www.edf.org/media/statement-environmental-defense-fund-president -fred-krupp-recent-events-baton-rouge.

Kurlander, Eric. "Hitler's Monsters: The Occult Roots of Nazism and the Emergence of the Nazi 'Supernatural Imaginary.'" *German History* 30, no. 4 (2012): 528–49.

Lacan, Jacques. *The Seminar of Jacques Lacan: Book I, Freud's Papers on Technique, 1953–1954*. Ed. Jacques-Alain Miller. New York: Norton, 1991.

——. *The Seminar of Jacques Lacan: Book III, The Psychoses, 1955–1956*. Ed. Jacques-Alain Miller. Trans. Russell Grigg. New York: Norton, 1997.

Larco, Nico. "Untapped Density: Site Design and the Proliferation of Suburban Multifamily Housing." *Journal of Urbanism: International Research on Placemaking and Urban Sustainability* 2, no. 2 (2009): 167–86.

Lazar, Nomi Claire. *States of Emergency in Liberal Democracies*. Cambridge: Cambridge University Press, 2009.

Lazarus, Richard. "Environmental Racism! That's What It Is." *University of Illinois Law Review* 2000, no. 1 (2000): 255–74.

Lee Myung-bak. "A Great People with New Dreams." *Presidential Speeches*, August 15, 2008. Accessed June 8, 2014. http://www.korea.net/Government/Briefing-Room/Presidential-Speeches/view?articleId=91000&pageIndex=9.

Lee, Robert G., and Sabine Wilke. "Forest as *Volk: Ewiger Wald* and the Religion of Nature in the Third Reich." *Journal of Social and Ecological Boundaries* 1, no. 1 (spring 2005): 21–46.

Lefebvre, Henri. *The Production of Space*. Trans. Donald Nicholson-Smith. Oxford: Wiley-Blackwell, 1992.

——. *The Urban Revolution*. Trans. Robert Bononno. Minneapolis: University of Minnesota Press, 2003.

Leonard, Annie. *The Story of Stuff: How Our Obsession with Stuff Is Trashing the Planet, Our Communities, and Our Health*. New York: Free Press, 2010.

Lewis, Paul, and Rob Evans. "Mark Kennedy: A Journey from Undercover Cop to 'Bona Fide' Activist." *Guardian*, January 10, 2011. Accessed December 16, 2014. http://www.theguardian.com/environment/2011/jan/10/mark-kennedy-undercover-cop-activist.

Liebes, Tamar, and James Curran, eds. *Media, Ritual and Identity*. London: Routledge, 1998.

Lin, Boqiang, and Chuanwang Sun. "Evaluating Carbon Dioxide Emissions in International Trade of China." *Energy Policy* 38, no. 1 (2010): 613–21.

Lin, Jintai, Da Pan, Steven J. Davis, Qiang Zhang, Kebin He, Can Wang, David G. Streets, Donald J. Wuebbles, and Dabo Guan. "China's International Trade and Air Pollution in the United States." *PNAS* 111, no. 5 (February 4, 2014): 1736–41. Accessed January 4, 2015. http://www.pnas.org/content/early/2014/01/16/1312860111.full.pdf.

Logie, David. "On the Frontlines of the Refugee Crisis." *Greenpeace*, April 3, 2016. Accessed July 17, 2015. http://www.greenpeace.org/usa/greenpeace-european-refugee-crisis-me/.

Lovelock, James. *Gaia: A New Look at Life on Earth*. Oxford: Oxford University Press, 1995.

Macklin, Graham D. "Co-opting the Counter Culture: Troy Southgate and the National Revolutionary Faction." *Patterns of Prejudice* 39, no. 3 (2005): 301–26.

Marx, Karl. *Capital: Volume 1.* Trans. Ben Fowkes. London: Penguin, 1990.

Matthäus, Jürgen, and Frank Bajohr. *The Political Diary of Alfred Rosenberg and the Onset of the Holocaust.* Lanham, Md.: Rowman and Littlefield, 2015.

Mayer, Sophie. "The Legend of Greenham Common Women's Peace Camp." *Transformation*, February 2, 2016. Accessed July 20, 2016. https://www.opendemocracy.net/transformation/sophie-mayer/bring ing-home-legend-of-greenham-common-womens-peace-camp.

McCarthy, John, and Mayer Zald. *The Trend of Social Movements in America: Professionalization and Resource Mobilization.* Morristown, N.J.: General Learning Press, 1973.

McDonough, William, and Michael Braungart. *Cradle to Cradle: Remaking the Way We Make Things.* New York: North Point Press, 2002.

Mendieta, Eduardo. "The Legal Orthopedia of Human Dignity: Thinking with Axel Honneth." *Philosophy and Social Criticism* 40, no. 8 (2014): 799–815.

Merrifield, Andy. "The Urban Question under Planetary Urbanization." *International Journal of Urban and Regional Research* 37, no. 3 (May 2013): 909–22.

Millennium Development Goals Indicators. Accessed February 15, 2016. http://mdgs.un.org/unsd/mdg/Metadata.aspx?IndicatorId=0&Series Id=711.

Miller, Matthew, and Peter Newcomb. "The World's 200 Richest People." *Bloomberg*, November 8, 2012. Accessed January 8, 2014. http://www .bloomberg.com/news/2012-11-01/the-world-s-200-richest-people.html.

Mirowski, Philip. *Never Let a Serious Crisis Go to Waste.* London: Verso, 2014.

Misselwitz, Philipp. "Shrinking Cities: Manchester/Liverpool." Working Papers, March 2004, 115–18. Accessed March 1, 2013. http://www .shrinkingcities.com/fileadmin/shrink/downloads/pdfs/WP-II_Man chester_Liverpool.pdf.

Moon, Ban Ki. *The Road to Dignity by 2030: Ending Poverty, Transforming All Lives and Protecting the Planet*, December 2014. Accessed December 26, 2015. http://www.un.org/ga/search/view_doc.asp?symbol=A/69/700&Lang=E.

Moore, Solomon. "As Program Moves Poor to Suburbs, Tensions Follow." *New York Times*, August 8, 2008. Accessed March 16, 2011. http://www .nytimes.com/2008/08/09/us/09housing.html?pagewanted=all&_r=0.

Mosse, George L. *The Crisis of German Ideology: Intellectual Origins of the Third Reich*. New York: Grosset and Dunlap, 1964.

Mouw, Ted. "Job Relocation and the Racial Gap in Unemployment in Detroit and Chicago, 1980 to 1990." *American Sociological Review* 65, no. 5 (2000): 730–53.

National Aeronautics and Space Administration. "Arctic Sea Ice Hits Smallest Extent in Satellite Era," September 16, 2012. Accessed April 1, 2014. http://www.nasa.gov/topics/earth/features/2012-seaicemin.html.

National-Anarchist Movement. "Part 5: Racial Separatism or Mixed Tribes?" September 18, 2010. Accessed July 21, 2016. http://www.national-anarchist .net/2010/09/part-5-racial-separatism.html.

——. "Part 9: Defence," September 18, 2010. Accessed July 21, 2016. http:// www.national-anarchist.net/2010/09/part-9-defence.html.

National Geographic Society. "Rain Forest Threats." Accessed August 11, 2016. http://environment.nationalgeographic.com/environment/habitats /rainforest-threats/.

National Oceanic and Atmospheric Administration. *Up-to-Date Weekly Average CO_2 at Mauna Loa*, 2013. Accessed June 1, 2014. http://www.esrl .noaa.gov/gmd/ccgg/trends/weekly.html.

National Resources Defense Fund. "The BP Oil Disaster at One Year." *National Resources Defense Fund*, 2011. Accessed April 2, 2014. http:// www.nrdc.org/energy/bpoildisasteroneyear.asp.

Neuwirth, Robert. *Shadow Cities: A Billion Squatters, a New Urban World*. New York: Routledge 2006.

Nova, Fritz. *Alfred Rosenberg: Nazi Theorist of the Holocaust*. New York: Hippocrene, 1986.

O'Connor, James. *Natural Causes: Essays in Ecological Marxism*. New York: Guilford Press, 1998.

OpenSecrets, Center for Responsive Politics. "Lobbying Database," 2014. Accessed June 1, 2014. http://www.opensecrets.org/lobby/.

Organisation for Economic Co-operation and Development. *Moving towards a Common Approach on Green Growth Indicators: A Green Growth Knowledge Platform*, 2013. Accessed May 10, 2014. http://issuu.com/ ggkp/docs/ggkp_moving_towards_a_common_approa.

——. *Towards Green Growth*, 2011. Accessed May 10, 2014. http://www.oecd .org/dataoecd/37/34/48224539.pdf.

Orlov, Oleg. "Ukraine's Forgotten City Destroyed by War." *Guardian*, January 7, 2015. Accessed August 2, 2016. https://www.theguardian.com/world/2015/jan/07/-sp-ukraine-pervomaisk-luhansk-forgotten-city-destroyed-by-war.

Orr, David. *The Nature of Design: Ecology, Culture, and Human Intention.* Oxford: Oxford University Press, 2002.

Osborne, Hilary. "Repossessions at Highest Level Since 1995." *Guardian*, February 11, 2010. Accessed December 21, 2012. http://www.theguardian.com/money/2010/feb/11/home-repossessions-highest-level-1995.

Ostrom, Elinor. *Governing the Commons: The Evolution of Institutions for Collective Action.* Cambridge: Cambridge University Press, 1990.

Oswalt, Philipp, and Tim Rieniets, eds. *Atlas of Shrinking Cities.* Stuttgart: Hatje Cantz, 2006.

Owen, David. *Green Metropolis: Why Living Smaller, Living Closer, and Driving Less Are the Keys to Sustainability.* New York: Riverhead Books, 2009.

Owusu, Henry. "Current Convenience, Desperate Deforestation: Ghana's Adjustment Program and the Forestry Sector." *Professional Geographer* 50, no. 4 (November 1998): 418–36.

Oxfam. *Oxfam Media Briefing* (ref. 02/2013), January 18, 2013. Accessed January 8, 2014. http://www.oxfam.org/sites/www.oxfam.org/files/cost-of-inequality-oxfam-mb180113.pdf.

——. "62 People Own Same as Half World." *Press Releases*, January 18, 2016. Accessed July 1, 2016. http://www.oxfam.org.uk/media-centre/press-releases/2016/01/62-people-own-same-as-half-world-says-oxfam-inequality-report-davos-world-economic-forum.

Pacheco, Tânia. "Inequality, Environmental Injustice, and Racism in Brazil: Beyond the Question of Colour." *Development in Practice* 18, no. 6 (November 2008): 713–25.

Palleroni, Sergio, and Christina Merkelback. *Studio at Large: Architecture in Service of Global Communities.* Seattle: University of Washington Press, 2004.

Palosaari, Marika. "Environmental Security." *United Nations Environment Program.* Accessed January 2, 2015. http://www.unep.org/roe/KeyActivities/EnvironmentalSecurity/tabid/54360/Default.aspx.

Pardo, Steve. "140 Acres in Detroit Sold to Grow Trees." *Detroit News*, December 12, 2012. Accessed December 14, 2012. http://www.detroitnews.com/article/20121212/METRO01/212120340.

"Paris Terror Attacks: Newspaper Front Pages from Around the World, in Pictures." Accessed October 23, 2016. http://www.telegraph.co.uk/news /worldnews/europe/france/11995738/Paris-Terror-Attacks-Newspaper -front-pages-from-around-the-world-in-pictures.html?frame=3500717.

Parr, Adrian. *Hijacking Sustainability*. Cambridge, Mass.: MIT Press, 2009.

——. *The Wrath of Capital: Neoliberalism and Climate Change Politics*. New York: Columbia University Press, 2013.

Parr, Adrian, and Natasha Leonard. "Our Crime Against the Planet and Ourselves." *New York Times*, May 18, 2016. Accessed July 1, 2016. http: //www.nytimes.com/2016/05/18/opinion/our-crime-against-the-planet -and-ourselves.html?_r=0.

Parr, Adrian, and Michael Zaretsky, eds. *New Directions in Sustainable Design*. Oxon: Routledge, 2011.

Patel, Raj. *Stuffed and Starved: The Hidden Battle for the World Food System*. New York: Marble House, 2008.

Pearlman, Jonathan. "100,000 Bats Fall Dead from the Sky During a Heat-wave in Australia." *Daily Telegraph*, January 8, 2014. Accessed March 1, 2014. http://www.telegraph.co.uk/news/worldnews/australiaandthepa cific/australia/10558183/100000-bats-fall-dead-from-the-sky-during-a -heatwave-in-Australia.html.

Peck, Jamie. *Constructions of Neoliberal Reason*. Oxford: Oxford University Press, 2010.

Pietikäinen, Petteri. "The Volk and Its Unconscious: Jung, Hauer, and the German Revolution." *Journal of Contemporary History* 35, no. 4 (October 2000): 523–39.

Piketty, Thomas. *Capital in the Twenty-First Century*. Trans. Arthur Gold-hammer. Cambridge, Mass.: Belknap Press, 2014.

Pois, Robert. *National Socialism and the Religion of Nature*. New York: St. Martin's Press, 1986.

Public Broadcasting Commission. *Frontline: Power Politics*. Accessed June 10, 2014. http://www.pbs.org/wgbh/pages/frontline/shows/blackout/ traders/power.html.

Putnam, Robert D. "Bowling Alone: America's Declining Social Capital." *Journal of Democracy* 6, no. 1 (January 1995): 65–78.

——. *Bowling Alone: The Collapse and Revival of American Community*. New York: Simon and Schuster, 2000.

——. *Making Democracy Work: Civic Traditions in Modern Italy.* Princeton, N.J.: Princeton University Press, 1993.

Rancière, Jacques. *The Emancipated Spectator.* Trans. Gregory Elliott. London: Verso, 2009.

——. *Figures of History.* Trans. Julie Rose. Cambridge: Polity Press, 2014.

——. *The Politics of Aesthetics.* Ed. and trans. Gabriel Rockhill. London: Continuum, 2004.

Rankin, Katherine N. "Social Capital, Microfinance, and the Politics of Development." *Feminist Economics* 8, no. 1 (2002): 1–24.

Ravallion, Martin. "How Long Will It Take to Lift One Billion People Out of Poverty?" *World Bank Research Observer* 28, no. 2 (August 2013): 139–58.

RealtyTrac. "Foreclosure Activity Increases 4 Percent in July," August 12, 2010. Accessed March 3, 2011. http://www.realtytrac.com/content/press-releases/foreclosure-activity-increases-4-percent-in-july-5946.

Reich, Wilhelm. *The Mass Psychology of Fascism.* Ed. Mary Higgins and Chester M. Raphael New York: Farrar, Straus and Giroux, 1970.

Retort. *Afflicted Powers: Capital and Spectacle in a New Age of War.* London: Verso, 2005.

Reynolds, Paul. "Kyoto: Why Did the US Pull Out?" *BBC News*, March 30, 2001. Accessed January 3, 2014. http://news.bbc.co.uk/2/hi/americas/1248757.stm.

Richardson, John, ed. *Handbook of Theory and Research for the Sociology of Education.* New York: Greenwood Press, 1986.

Riehl, Wilhelm Heinrich. "Field and Forest" (1853). Trans. Frances H. King. In *The German Classics: Volume VIII* (1913), ed. Kuno Francke. Accessed July 22, 2016. http://www.unz.org/Pub/FranckeKuno-1913v08-00410.

Robine, Jean-Marie, Siu Lan K. Cheung, Sophie Le Roy, Herman Van Oyen, Clare Griffiths, Jean-Pierre Michel, and François Richard Herrmann. "Death Toll Exceeded 70,000 in Europe During the Summer of 2003." *Comptes rendus biologies* 331, no. 2 (2008): 171–78.

Rosen, Michael. *Dignity.* Cambridge, Mass.: Harvard University Press, 2012.

Rossi, Aldo. *The Architecture of the City.* Trans. Diane Ghirardo and Joan Ockman. Cambridge, Mass.: MIT Press, 1982.

Roubini, Nouriel, and Stephen Mihm. *Crisis Economics: A Crash Course in the Future of Finance*. New York: Penguin, 2010.

Roy, Ananya, Genevieve Negrón-Gonzales, Kweku Opoku-Agyemang, and Clare Talwalker. *Encountering Poverty: Thinking and Acting in an Unequal World*. Berkeley: University of California Press, 2016.

Roy, Arundhati. *The Cost of Living*. Toronto: Modern Library, 1999.

Saad, Lydia, and Jeffrey M. Jones. "U.S. Concern About Global Warming at Eight-Year High." *Gallup: Politics*, March 16, 2016. Accessed June 30, 2016. http://www.gallup.com/poll/190010/concern-global-warming -eight-year-high.aspx?g_source=CATEGORY_CLIMATE_CHANGE &g_medium=topic&g_campaign=tiles.

Said, Edward. "On Dignity and Solidarity." Lecture, Washington, D.C., June 15, 2003. *Democracy Now*. Accessed February 11, 2016. http://www .democracynow.org/2003/10/20/on_dignity_and_solidarity_scholar _activist.

Scarry, Elaine. *Thinking in an Emergency*. New York: Norton, 2011.

Schlichting, Inga. "Strategic Framing of Climate Change by Industry Actors: A Meta-Analysis." *Environmental Communication* 7, no. 4 (December 2013): 493–511.

Schmitt, Carl. *Political Theology: Four Chapters on the Concept of Sovereignty*. Trans. George Schwab. Cambridge, Mass.: MIT Press, 1985.

Sharp, Gene. *Gandhi as a Political Strategist*. Boston: Porter Sargent, 1979.

Shearer, Richard. "The Small Town–Big City Split That Elected Donald Trump." Brookings Institution, November 11, 2016. Accessed March 18, 2017. https://www.brookings.edu/blog/the-avenue/2016/11/11/the-small -town-big-city-split-that-elected-donald-trump/.

Smith, Maf, John Whitelegg, and Nick Williams. *Greening the Built Environment*. London: Routledge, 1998.

Shershow, Scott Cutler. *Deconstructing Dignity: A Critique of the Right-to-Die Debate*. Chicago: University of Chicago Press, 2014.

Smyth, Sharon, and Charles Penty. "Spain Foreclosures Spread to Once Wealthy." *Bloomberg*, October 9, 2012. Accessed December 7, 2012. http://www.bloomberg.com/news/articles/2012-10-08/spain-foreclosures -spread-to-once-wealthy-mortgages.

Snyder, Scott A. "Assessing the Global Green Growth Institute (GGGI) and the Sustainability of South Korea's Contribution." *Council on Foreign Relations*, November 6, 2012. Accessed December 1, 2014. http://blogs.cfr

.org/asia/2012/11/06/assessing-the-global-green-growth-institute-gggi
-and-the-sustainability-of-south-koreas-contribution/.

Soja, Edward W. *Postmodern Geographies: The Reassertion of Space in Critical Social Theory*. London: Verso, 2011.

Sontag, Susan. "The Imagination of Disaster." *Commentary*, October 1 1965, 42–48.

Southgate, Troy. "Transcending the Beyond: Third Position to National-Anarchism." *Pravda*, January 17, 2002. Accessed July 21, 2016. http://www.pravdareport.com/news/russia/17-01-2002/25940-0/.

Spinoza, Benedict de. *Ethics*. Ed. and trans. Edwin Curley. London: Penguin, 1996.

Staggenborg, Suzanne. "Coalition Work in the Pro-Choice Movement: Organizational and Environmental Opportunities and Obstacles." *Social Problems* 33, no. 5 (June 1986): 374–90.

Standing, André, and Michael Gachanja. *A Corruption Risk Assessment for REDD+ in Kenya*. Ministry of Environment, Water and Natural Resources and UN-REDD Programme, 2013. Accessed December 3, 2014. http://www.kenyaforestservice.org/documents/redd/Analytical%20 Study%20on%20Corruption%20Risk%20Assessment%20for%20 REDD+%20in%20Kenya.pdf.

Steffen, Alex. *Carbon Zero: Imagining Cities That Can Save the Planet*. Mountain View, Calif.: Creative Commons, 2012.

Stern, Nicholas. *The Economics of Climate Change: The Stern Review*. Cambridge: Cambridge University Press, 2007.

Stiglitz, Joseph E. *Free Fall: America, Free Markets, and the Sinking of the World Economy*. New York: Norton, 2010.

Sunshine, Spencer. "Rebranding Fascism: National-Anarchists." *Public Eye* 23, no. 4 (winter 2008). Accessed July 21, 2016. http://www.publiceye.org/magazine/v23n4/rebranding_fascism.html.

Surin, Kenneth. *Freedom Not Yet: Liberation and the Next World Order*. Durham, N.C.: Duke University Press, 2009.

Suzuki, David. "Carbon Offsets as a Tool in the Fight Against Global Warming." *David Suzuki Foundation*, 2009. Accessed April 10, 2014. http://www.straight.com/news/david-suzuki-carbon-offsets-tool-fight -against-global-warming.

Swift, Anthony. "How an Unlikely Coalition of Environmental Activists Stopped Keystone XL." *Huffington Post*, November 9, 2015.

Accessed July 18, 2016. http://www.huffingtonpost.com/anthony-swift /unlikely-environmental-activists-stopped-keystone-xl_b_8513650 .html.

Syrjänen, Raakel. *UN-HABITAT and the Kenya Slum Upgrading Programme: Strategy Document.* UN-HABITAT, 2008. Accessed February 11, 2016 .https://unhabitat.org/books/un-habitat-and-the-kenya-slum-upgrading -programme-strategy-document/.

350.org. "Stop the Keystone XL Pipeline." Accessed July 18, 2016. http://350 .org/campaigns/stop-keystone-xl/.

Tidwell, Thomas. "Wildland Fire Management." *Statement before the Committee on Energy and Natural Resources, U.S. Senate,* June 4, 2013. Accessed March 1, 2014. http://www.energy.senate.gov/public/index.cfm/files/ serve?File_id=e59df65c-09c6-4ffd-9a83-f61f2822a075.

Tilman, David, Robert M. May, Clarence L. Lehman, and Martin A. Nowak. "Habitat Destruction and the Extinction Debt." *Nature* 371 (September 1994): 65–66.

Torgerson, Douglas. *The Promise of Green Politics: Environmentalism and the Public Sphere.* Durham, N.C.: Duke University Press, 1999.

Uekoetter, Frank. *The Green and the Brown: A History of Conservation in Nazi Germany.* Cambridge: Cambridge University Press, 2006.

UNEP, UNDP, NATO, OSCE. *Environment and Security: Transforming Risks into Cooperation,* 2005. Accessed January 2, 2015. http://www.envsec .org/publications/ENVSEC.Transforming%20risks%20into%20coop eration.%20Central%20Asia.%20Ferghana-Osh-Khujand%20area_ English.pdf.

UNHCR. *Global Trends: Forced Displacement in 2015.* Accessed August 9, 2016. https://s3.amazonaws.com/unhcrsharedmedia/2016/2016-06-20 -global-trends/2016-06-14-Global-Trends-2015.pdf.

——. "Worldwide Displacement Hits All-Time High as War and Persecution Increase," June 18, 2015. Accessed August 5, 2016. http://www.unhcr .org/en-us/news/latest/2015/6/558193896/worldwide-displacement-hits -all-time-high-war-persecution-increase.html.

United Nations. *Reducing Emissions That Cause Climate Change.* Gateway to the UN System's Work on Climate Change, 2013. Accessed May 14, 2014. http://www.un.org/wcm/content/site/climatechange/pages/gate way/mitigation/reducing-emissions.

——. "Universal Declaration of Human Rights." Accessed January 15, 2016. http://www.un.org/en/universal-declaration-human-rights/.

——. *Urban Millennium*, June 6–8, 2001. Accessed August 8, 2016. http://www.un.org/ga/Istanbul+5/booklet4.pdf.

——. "We Can End Poverty: Millennium Development Goals and Beyond 2015." *UN Millennium Goals*. Accessed January 21, 2016. http://www.un.org/millenniumgoals/environ.shtml.

——. *World Economic and Social Survey 2013: Sustainable Development Challenges*. Accessed February 6, 2014. http://sustainabledevelopment.un.org/content/documents/2843WESS2013.pdf.

——. "World's Population Increasingly Urban with More Than Half Living in Urban Areas," July 10, 2014. Accessed August 8, 2016. http://www.un.org/en/development/desa/news/population/world-urbanization-prospects-2014.html.

United Nations Central Emergency Response Fund. "Philippines: UN Releases US$25 Million to Fund Emergency Response," November 11, 2013. Accessed December 20, 2014. http://www.unocha.org/cerf/resources/top-stories/philippines-un-releases-us25-million-fund-emergency-response.

United Nations Conference on Trade and Development. "Report on UNCTAD Assistance to the Palestinian People: Developments in the Economy of the Occupied Palestinian Territory," July 6, 2015. Accessed August 9, 2016. http://unctad.org/en/PublicationsLibrary/tdb62d3_en.pdf.

United Nations Educational, Scientific and Cultural Organization. "Director-General of UNESCO Irina Bokova Firmly Condemns the Destruction of Palmyra's Ancient Temple of Baashamin, Syria." *UNESCO News*, August 24, 2015. Accessed August 11, 2016. http://whc.unesco.org/en/news/1339/.

——. *Site of Palmyra*. Accessed August 11, 2016. http://whc.unesco.org/en/list/23.

United Nations Environment Program. *Green Economy Report*, 2010. Accessed May 22, 2014. http://www.unep.org/greeneconomy/.

——. *Kenya: Atlas Of Our Changing Environment*, 2009. Accessed March 17, 2017. http://staging.unep.org/pdf/Kenya_Atlas_Full_EN_72dpi.pdf.

United Nations Radio. "More Than 1.5 Billion People Still Live in Conflict-Affected Countries," February 27, 2013. Accessed August 8, 2016.

http://www.unmultimedia.org/radio/english/2013/02/more-than-1-5
-billion-people-still-live-in-conflict-affected-countries-escap/.

United States Census Bureau. "2010 Census Urban Area FAQs: Urban-Rural Classification Program." Accessed August 8, 2016. https://www.census.gov/geo/reference/ua/uafaq.html.

United States Conference of Catholic Bishops. "Renewing the Earth," November 14, 1991. Accessed February 2, 2016. http://www.usccb.org/issues-and-action/human-life-and-dignity/environment/renewing-the-earth.cfm.

United States Environmental Protection Agency. "Environmental Justice Program and Civil Rights," 2014. Accessed October 10, 2014. http://www.epa.gov/region1/ej/.

Van Natta, Don, Jr. "Enron's Collapse: The Politicians; Enron Spread Contributions on Both Sides of the Aisle." *New York Times*, January 21, 2002.

Vattimo, Gianni, and Santiago Zabala. *Hermeneutic Communism: From Heidegger to Marx*. New York: Columbia University Press, 2011.

Vidal, John. "Revealed: How Oil Giant Influenced Bush." *Guardian*, June 8, 2005. Accessed December 20, 2013. http://www.theguardian.com/news/2005/jun/08/usnews.climatechange.

——. "Water Supply Key to Outcome of Conflicts in Iraq and Syria, Experts Warn." *Guardian*, July 2, 2014. Accessed January 2, 2015. http://www.theguardian.com/environment/2014/jul/02/water-key-conflict-iraq-syria-isis.

Virilio, Paul, Diller Scofidio + Renfro, Mark Hansen, Laura Kurgan, and Ben Rubin, with Robert Gerard Pietrusko, and Stewart Smith. *EXIT* (2008–2015). https://www.youtube.com/watch?v=kyMbF2uuSIw.

Voruz, Véronique, and Bogdan Wolf, eds. *The Later Lacan: An Introduction*. Albany: SUNY Press, 2007.

Weber, Christopher, Glen Peters, Dabo Guan, and Klaus Hubacek. "The Contribution of Chinese Exports to Climate Change." *Energy Policy* 36, no. 9 (2008): 3572–77.

Welch, Christina. "The Spirituality of, and at, Greenham Common Peace Camp." *Feminist Theology* 18, no. 2 (2010): 230–48.

The White House, Office of the Press Secretary. "President Obama Signs Washington Emergency Declaration," March 24, 2014. Accessed December 20, 2014. http://www.whitehouse.gov/the-press-office/2014/03/24/president-obama-signs-washington-emergency-declaration.

——. "Remarks by the President at the National Defense University," May 23, 2013. Accessed August 2, 2016. https://www.whitehouse.gov/the-press-office/2013/05/23/remarks-president-national-defense-university.

Wike, Richard. "What the World Thinks About Climate Change in 7 Charts." *Pew Research Center*, April 18, 2016. Accessed July 1, 2016. http://www.pewresearch.org/fact-tank/2016/04/18/what-the-world-thinks-about-climate-change-in-7-charts/.

Wilmes, Adam R. *Altruism by Design: How to Effect Social Change as an Architect.* London: Routledge, 2015.

Withnall, Adam. "Global Peace Index 2016: There Are Now Only 10 Countries in the World That Are Actually Free from Conflict." *Independent*, June 8, 2016. Accessed August 11, 2016. http://www.independent.co.uk/news/world/politics/global-peace-index-2016-there-are-now-only-10-countries-in-the-world-that-are-not-at-war-a7069816.html.

Wolf, Marcus, Kenneth Haar, and Olivier Hoedeman. *The Fire Power of the Financial Lobby: A Survey of the Size of the Financial Lobby at the EU Level*, April 2014. Accessed June 6, 2014. http://corporateeurope.org/sites/default/files/attachments/financial_lobby_report.pdf.

Wong, Vanessa. "Goodbye, Renters Market: Rents Grow as Demand Increases." *Bloomberg Business Week*, March 21, 2011. Accessed December 12, 2012. http://www.realestate.msn.com/article.aspx?cp-documentid=28001698.

World Bank. "About the World Bank" Accessed February 1, 2016. http://www.worldbank.org/en/about.

——. "Extreme Poverty Rates Continue to Fall," *World Bank*, June 2, 2010. Accessed February 1, 2016. http://blogs.worldbank.org/opendata/extreme-poverty-rates-continue-fall.

——. *Global Economic Prospects 2007: Managing the Next Wave of Globalization.* Washington, D.C.: International Bank for Reconstruction and Development/World Bank, 2007.

——. *Inclusive Green Growth: The Pathway to Sustainable Development*, 2012. Accessed May 10, 2014. http://issuu.com/world.bank.publications/docs/9780821395516.

——. *World Development Report 2009: Reshaping Economic Geography.* Washington, D.C.: World Bank, 2008.

Zax, Jeffrey S., and John F. Kain. "Moving to the Suburbs: Do Relocating Companies Leave Their Black Employees Behind?" *Journal of Labor Economics* 14, no. 3 (1996): 472–504.

Zerilli, Linda M. G. "'We Feel Our Freedom': Imagination and Judgment in the Thought of Hannah Arendt." *Political Theory* 33, no. 2 (April 2005): 158–88.

Žižek, Slavoj. *Living in the End Times*. London: Verso, 2011.

——. *Looking Awry: An Introduction to Jacques Lacan Through Popular Culture*. Cambridge, Mass.: MIT Press, 1992.

——. *Organs without Bodies*. Oxon: Routledge, 2004.

——. *The Parallax View*. Cambridge, Mass.: MIT Press, 2006.

——. *The Sublime Object of Ideology*. London: Verso, 1989.

——. *Trouble in Paradise: From the End of History to the End of Capitalism*. London: Allen Lane, 2014.

——. *Violence: Six Sideways Reflections*. New York: Picador, 2008.

——. *Welcome to the Desert of the Real*. London: Verso, 2002.

Zupančič, Alenka. *Ethics of the Real: Kant and Lacan*. London: Verso, 2000.

INDEX

GPSR Authorized Representative: Easy Access System Europe, Mustamäe tee 50, 10621 Tallinn, Estonia, gpsr.requests@easproject.com

www.ingramcontent.com/pod-product-compliance
Lightning Source LLC
Chambersburg PA
CBHW022137020426
42334CB00015B/942